Net Loss:
Fish, Jobs, and the
Marine Environment

PETER WEBER

Anne Platt, *Staff Researcher*

Carole Douglis, *Editor*

WORLDWATCH PAPER 120
July 1994

THE WORLDWATCH INSTITUTE is an independent, nonprofit environmental research organization based in Washington, D.C. Its mission is to foster a sustainable society—in which human needs are met in ways that do not threaten the health of the natural environment or future generations. To this end, the Institute conducts interdisciplinary research on emerging global issues, the results of which are published and disseminated to decisionmakers and the media.

FINANCIAL SUPPORT is provided by the Geraldine R. Dodge Foundation, W. Alton Jones Foundation, John D. and Catherine T. MacArthur Foundation, Andrew W. Mellon Foundation, Curtis and Edith Munson Foundation, Edward John Noble Foundation, Pew Charitable Trusts, Lynn R. and Karl E. Prickett Fund, Rockefeller Brothers Fund, Surdna Foundation, Turner Foundation, Frank Weeden Foundation, Wallace Genetic Foundation, and Peter Buckley.

PUBLICATIONS of the Institute include the annual *State of the World*, which is now published in 27 languages; *Vital Signs*, an annual compendium of the global trends—environmental, economic, and social—that are shaping our future; the *Environmental Alert* book series; and *World Watch* magazine, as well as the *Worldwatch Papers*. For more information on Worldwatch publications, write: Worldwatch Institute, 1776 Massachusetts Ave., N.W., Washington, DC 20036; or FAX (202) 296-7365.

THE WORLDWATCH PAPERS provide in-depth, quantitative and qualitative analysis of the major issues affecting prospects for a sustainable society. The Papers are authored by members of the Worldwatch Institute research staff and reviewed by experts in the field. Published in five languages, they have been used as a concise and authoritative reference by governments, nongovernmental organizations and educational institutions worldwide. For a partial list of available Papers, see page 77.

DATA from all graphs and tables contained in this book, as well as from those in all other Worldwatch publications of the past year, are available on diskette for use with Macintosh or IBM-compatible computers. This includes data from the *State of the World* series, *Vital Signs* series, Worldwatch Papers, *World Watch* magazine, and the *Environmental Alert* series. The data are formatted for use with spreadsheet software compatible with Lotus 1-2-3, including Quattro Pro, Excel, SuperCalc, and many others. Both 3 1/2" and 5 1/4" diskettes are supplied. To order, send check or money order for $89, or credit card number and expiration date (Visa and MasterCard only), to Worldwatch Institute, 1776 Massachusetts Ave., NW, Washington, DC 20036. Tel: 202-452-1999; Fax: 202-296-7365; Internet: wwpub@igc.apc.org.

Table of Contents

Tables and Figures

ACKNOWLEDGMENTS: Innumerable people and organizations made this study possible. The Curtis and Edith Munson Foundation provided specific support for this paper. I would also like to thank the staff of the FAO Fisheries Department, who provided much of the data upon which this study is based. For their timely review and input, I would like to thank John Caddy, Mac Chapin, Adele Crispoldi, Vlad Kaczynski, Eduardo Loayza, Alair MacLean, Chris Newton, Lisa Speer, Mike Weber, and John Wise.

My colleagues at Worldwatch deserve a special thanks: Anne Platt for her energetic and insightful research, Sandra Postel for her steady guidance and direction, and Carole Douglis for immersing herself in the editing and refining of the text. Finally, I would like to acknowledge the people doing field research, only a fraction of whom I was even able to cite.

PETER WEBER is a Research Associate at the Worldwatch Institute, and coauthor of two of the Institute's *State of the World* reports. He has also written Worldwatch Paper 116, *Abandoned Seas: Reversing the Decline of the Oceans.*

Introduction

On April 2, 1994, Canadian enforcement officers wielding machine guns boarded and seized a foreign fishing vessel 45 kilometers outside Canada's jurisdiction in the North Atlantic Ocean. Although the seizure violated international law, in Canada's view the boat's crew—Portuguese nationals flying a Panamanian flag—were modern-day pirates stealing Canada's cod and keeping 30,000 Canadian fishers out of work.[1]

Far from backing down in the face of international displeasure, the Canadian Parliament in May enacted a law unilaterally declaring Canada authorized to confiscate any foreign vessel working the Grand Banks fishing grounds outside Canada's internationally recognized 200-mile limit.[2]

Such gunboat diplomacy exemplifies the tensions brewing in the world's fishing grounds. Seizures of foreign vessels are only the most high profile incidents. Closer to shore, declining fish stocks have brought neighboring fishers to the edge of violence and sent shudders through coastal communities. One Maine lobster fisher found a pipe bomb in one of his traps; he was lucky, but others have died in fishing conflicts. Some fishers nervously watch as polluters set up shop next to their fishing grounds, like the Texas shrimp fishers who see petrochemical plants spread along the coast of the Gulf of Mexico. Fishers everywhere worry about the future as they make more effort for less catch—or are banned from their livelihood outright.[3]

The fundamental problem that fishers face is their own ability to catch fish and counterproductive government policies that have led more people and boats into the business even after the point of diminishing returns. After decades of buying bigger boats and more advanced hunting technologies, fishers have nearly fished the oceans to the limits. Of the world's 15 major marine fishing regions, the catch in all but 2 has fallen; in 4, the catch has shrunk by more than 30 percent. Since 1989, the marine catch of fish, crustaceans, such as lobster, and mollusks, such as clams, has stagnated. With fewer fish to net in many of the world's fishing grounds, fishers fear becoming scarcer than the fish they seek.[4]

Although worldwide environmental degradation of the oceans contributes to the decline of marine life, overfishing is the primary cause of dwindling fish populations. Unless fishery management policies change, we can expect a growing environmental and human toll. Catches will slip further, and millions of fishers will lose their jobs. "There is little reason to believe that the global catch can...expand, except for increases that might occur through more effective management of stocks," United Nations Food and Agriculture Organization (FAO) warns.[5]

Already declining catches have translated into job loss among the world's 15 to 21 million fishers. In the last few years, over 100,000 fishers around the world have lost their source of income. One hundred times that number could be out of work in the coming decades as countries try to come to grips with the great disparity between the capacity of the world's fishing fleets and the limits of the oceans. Worldwide, fishers possess on the order of twice the capacity needed to fish the oceans.[6]

Income generated from marine fishing fuels only a small part of the global economy, perhaps one percent. But in coastal and island regions, fishing takes on greater importance. In Southeast Asia, more than 5 million people fish full-time, contributing some $6.6 billion toward the region's aggregate national incomes. In northern Chile, fishing accounts for 40 percent of income, the employment of 18,000 people, and $400 million worth of exports in 1990. Iceland was founded on marine fishing, which today accounts for 17 percent of the national income and 12 to 13 percent of employment. Some 200 million people around the world depend on fishing and fish-related industries for their livelihoods.[7]

Tight times in marine fishing particularly harm coastal communities and cultures. From Canadian seaside villages to South Pacific island cultures, fishing is a social mainstay. Lost opportunities can turn fishing towns into ghost towns, and fishing cultures into lost cultures. Small-scale fishers—who get the least support from governments—form the backbone of community and cultural diversity along the world's coasts.

In developing countries, small-scale fishers are also the primary supplier of fish, particularly for local consumption. When fish enters the commercial market, it is no longer available to low-income consumers and subsistence cultures. Once considered the poor person's protein, fish is becoming expensive even for consumers in industrial countries, and some species that were once common in supermarkets are no longer readily available.[8]

Until the recent stagnation in the world catch, the supply of fish per person had been rising steadily. Marine fishing boomed after World War II, increasing seafood available for consumers around the world and far surpassing the catch from freshwater lakes and rivers. Today in Asia, an estimated one billion people rely on fish as their primary source of animal protein, as do many people in island nations and the coastal states of Africa. Worldwide, fish and other products of the sea account for 16 percent of animal protein consumption—more than either pork or beef—and 5.6 percent of total protein intake.[9]

With fewer fish to net in many of the world's fishing grounds, fishers fear becoming scarcer than the fish they seek.

However, distribution does not always follow need. An average resident of an industrial country consumes nearly three times as much fish as her counterpart in the developing world. In low-income countries such as Sierra Leone, where fish is the primary source of animal protein, consumption remains lower per person than in industrial North America and Europe. Meanwhile, low-income consumers are losing access to affordable fish as supplies tighten and high-priced markets attract a growing proportion of the world fish supply.[10]

Can the oceans continue to help meet the growing demand for food? The answer is two-fold. The oceans are not the unlimited reservoir of low-cost food they were once considered. But marine fisheries can help to reduce malnutrition and continue to support traditional coastal communities—if marine fishing is managed for this purpose.

Rather than contributing to a solution, however, government policies have for the most part promoted the overexpansion of the fishing industry. Worldwide, governments underwrite the growth of national fleets with subsidies on the order of $54 billion annually—to catch $70 billion worth of fish. Meanwhile, fishing grounds often remain open to all comers. The subsidies and open access policies have been particularly disruptive for traditional communities that had successful fisheries management systems in place before outside intervention.[11]

With catches declining and jobs imperiled in many fisheries, countries need to act. Fish, fishers and eaters would all benefit from programs to rehabilitate and protect marine fisheries. Fishers could potentially increase their catch by 20 million tons—25 percent of the current catch—if they allowed fish populations to rebuild.[12]

Finding the political will to change fishing policies, however, is harder than finding fish. The overcapacity of the world's fishing fleets means that the industry is in for a period of painful readjustment. Who gets squeezed out has enormous implications for jobs and coastal communities. Either the industrial fishing fleets or the community-based fishers are going pay a heavy price. If countries continue to favor large-scale, industrial style fishing, some 14-20 million small-scale fishers and their communities are at risk.[13]

While no changes will receive universal approval, the possibility of successful management improves when governments and communities cooperate. The combined effort can protect public resources while keeping the day-to-day decision-making at the local level. Without community-based control, marine fisheries will be depleted not only of fish, but also of the social benefits they have long provided.

Limits of the Sea

One of the factors that separated early humans from primates was the ability to gather food from the sea. In fact, archaeological evidence suggests that the first people to turn from nomadism to more settled ways were fishers, not farmers, as is commonly thought. The first marine fishers were the Maglemosians, who developed settlements in northern Europe some 10,000 years ago. Along the Baltic seacoast they left behind large piles of shellfish and smaller deposits of fish bones, as well as barbed hooks, harpoons, and primitive boats. The size and centralized design of the Maglemosian settlements indicate they had large food surpluses compared to their hunter-gatherer contemporaries. Other sites of early fishing cultures have been found at the mouth of the Nile and in Baja California, Japan, and Peru.[14]

Marine fisheries have sustained coastal communities for millennia, and the scale of the oceans has given people the impression that we could boost the catch for millennia to come. The term "fishery" can refer simultaneously to the people, equipment, species, and/or regions involved in fishing. Therefore, one can refer to marine or freshwater fisheries, commercial or traditional, cod, anchovy, large-scale or small-scale, coastal or high-seas—even whale fisheries. Fishers, however, are now running up against the limits of the marine fisheries. Most of the potential for increasing catches in the future lies in allowing overfished and depleted stocks to recuperate.

The first fishers relied on marine life, such as shellfish, that they could catch near shore. They benefitted from the fact that coastal ecosystems—estuaries, wetlands, and reefs—are among the most productive ecosystems on earth. With simple gear and boats, fishers can catch large quantities of seafood in coastal waters.

Coastal areas are still key to marine fisheries. Nutrients washing off land support a higher density of life than can be found in the open oceans. More than 90 percent of the catch comes from the 10 percent of the oceans closest to land. The major offshore fisheries lie where continental shelves extend far into the ocean, such as in the Grand Banks off the east coast of Canada, or where marine currents concentrate nutrients, such as the upwelling waters off Peru. Open ocean fishing is usually not worth the effort except when fishers can locate species that hunt in schools, including tuna—but these roam an area half the earth's surface.[15]

How much food can we pull from the oceans? The answer depends on what is considered worth catching. The primary production of the oceans is estimated to be on the order of 190 billion tons per year of microscopic phytoplankton and other marine plants. The next level of the food chain contains primarily zooplankton, bacteria, and viruses. Then come small fish, bigger fish, and so on.[16]

From one layer of the food chain to the next, the biomass decreases by approximately a factor of ten. Most of the species that people enjoy are three or four levels up the food chain. Therefore the level of production per year is 1/1,000 to 1/10,000 the primary production. Needless to say, people share these sea creatures with other predators, from sharks to birds.[17]

Based on this type of data, numerous studies have attempted to gauge the limits of the oceanic fish catch. In 1968, scientists at an international conference at the University of Washington estimated the range of possibilities to be between 80 million and 2 billion tons—suggesting 200 million tons as a realistic upper limit. Estimates since then have declined. In 1971, the FAO estimated that the marine environment could sustainably yield about 100 million tons of fish per year. In 1984 an FAO report on the potential of specific regions of the oceans calculated a similar yield. Although such estimates are inherently uncertain, the decline of major fisheries and the recent faltering of the world catch overall suggest that the more recent projections are useful guidelines.[18]

Today, fishery scientists use 100 million tons per year—about 20 million tons more than the 1993 marine catch—as a rough estimate of the potential for all commercially viable marine species in the foreseeable future. But these gains will not come easily: The marine catch is unlikely to reach and maintain the 100 million-ton mark unless fish stocks are better managed.[19]

In the fifties and sixties, the supply of fishery products grew at three times the rate of human population growth. Fishers seemed only to have to travel further and put out more nets, hooks, and traps if they wanted to catch more fish and other marine life. In this century, the marine catch rose more than 25-fold—from some 3 million tons at the turn of the century to a peak of 82 million tons in 1989. What is known as the "marine fish catch" consisted of 85 percent fish, 5 percent crustaceans, and 10 percent mollusks in 1991. The marine catch dwarfs the freshwater catch—only 6.4 million tons in 1989. (See Figure 1.)[20]

Yet in the post-war period, the growth in fish caught masked the depletion of particular sites and species. By the 1960s, the yields from major marine fisheries—including the centuries-old Atlantic cod fishery—were topping out and beginning to shrink. Only by reaching out to new fishing grounds and new species could fishers manage a net growth in supply. Fishing fleets rapidly expanded and modernized, progressively working their way from the most desirable, closest, or easy-to-catch species, to less desirable or more elusive ones.

The marine catch is unlikely to reach and maintain the 100-million-ton mark unless fish stocks are better managed.

In fact the explosive growth in the fish catch depended largely on increasing the haul of low-value species such as the Peruvian anchovy—a small, oily fish used primarily for animal feed. The Peruvian anchovy catch grew rapidly from its beginning in the 1950s to 13.1 million tons in 1970—becoming the world's single largest fish catch in the process.[21]

The era of expanding catches worldwide halted abruptly in the early seventies when a combination of overfishing and natural environmental change caused the Peruvian anchovy fishery to collapse. The Peruvian anchovy—1/5 the world's total fish haul at the time—tumbled from its peak of 13.1 million tons in 1970 to less than 2 million tons in 1974. A new era of limits had begun.[22]

FIGURE 1

World Fish Catch, 1950-1993

Million Tons

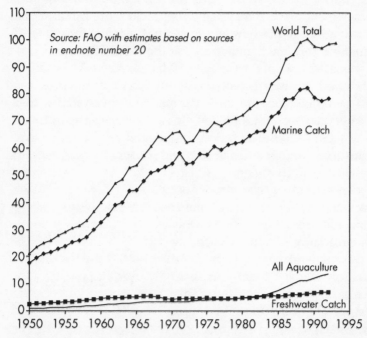

After this collapse, the world's fishing fleets could no longer sustain the growth trends of earlier in the century. Growth in the marine catch fell from six percent to around two percent per year in the 1970s and 1980s. Most of this two percent growth again came from low-value species such as the rebounding Peruvian anchovy fishery. The biggest gains since 1970—over 4,000,000 tons each—came from the sardine-like Japanese pilchard and the South American pilchard, whose catches totaled less than 20,000 tons per year two decades ago.[23]

Fishers have also gone to great lengths to increase their haul of high-value, hard-to-find species such as tuna and squid. Fishers use satellite data and aircraft to track tuna, and high-intensity lamps to congregate squid. In addition, fishers have employed bigger boats and gear, including driftnets, to strain large portions of the ocean. As a result, the catch of tuna and related species grew from 1.7 million tons in 1970 to 4.1 million tons in 1989; the catch of squid and related species (cephalopods) rose from 1 million tons to 2.7 million tons during the same period. Two species of tuna—the skipjack and the yellowfin—have contributed 700,000 and 500,000 tons respectively to the growth of the world fish catch between 1973 and 1989. Skipjack entered the top ten of the marine catch during the eighties.[24]

By the late eighties, however, fishers had tapped all the high-volume, low-value fisheries, and were pushing the limits of a number of the tuna and squid fisheries. Although some of the high-value and other minor fisheries continued to expand, major declines in other fisheries were off-setting these modest gains. In 1989 the world catch peaked; it has stagnated ever since.[25]

All the world's major fishing grounds are at or beyond their limits, and many have already suffered serious declines. Of the planet's 15 major marine fishing regions, the productivity in all but 2 has fallen. In four of the hardest hit areas—the northwest, west-central, and southeast Atlantic and the east-central Pacific—the total catch has shrunk by over 30 percent. (See Table 1.) With a long history of intensive fishing, the Atlantic fisheries have experienced the biggest drops, but the Mediterranean and parts of the Pacific have also seen large losses. Only the Indian Ocean fisheries are still increasing total output, although they are unlikely to expand much more and could be poised for some serious declines. The combined total decline from the peak years for the major marine fishing regions is 13 million tons.[26]

A look at the catch of individual species reveals even steeper declines. Eighteen fisheries have seen their productivity drop by more than 100,000 tons each. (See Table 2.) Together, these drops represent a fall of nearly 30 million tons—more than one-third of the 1992 marine catch.[27]

The range of declines from regional and individual fisheries—13 to 30 million tons—gives some indication not only of what has been lost, but of the potential for a rise in catch levels. FAO estimates that depleted fisheries could yield another 20 million tons annually from their current level—if fishers give fisheries time to recuperate. Fishers will not be able to regain the

TABLE 1

Change in Catch for Major Marine Fishing Regions, Peak Year to 1992

Region	Peak Year	Peak Catch	1992 Catch	Change
		(million tons)	(million tons)	(percent)
ATLANTIC OCEAN				
Northwest	1973	4.4	2.6	-42
Northeast*	1976	13.2	11.1	-16
West Central	1984	2.6	1.7	-36
East Central	1990	4.1	3.3	-20
Southwest	1987	2.4	2.1	-11
Southeast*	1973	3.1	1.5	-53
MEDITERRANEAN AND BLACK SEAS*	1988	2.1	1.6	-25
PACIFIC OCEAN				
Northwest	1988	26.4	23.8	-10
Northeast*	1987	3.4	3.1	- 9
West Central	1991	7.8	7.6	- 2
East Central	1981	1.9	1.3	-31
Southwest	1991	1.1	1.1	- 2
Southeast	1989	15.3	13.9	- 9
INDIAN OCEAN				
Western	still rising		3.7	+6**
Eastern	still rising		3.3	+5**

Source: FAO.

*Rebounding from a larger decline. **Average annual growth since 1988.

Note: Percentages calculated before rounding off catch figures. The catch in the Antarctic is at 356,000 tons, down from a peak of 653,000 tons in 1982, primarily because of reduced interest in krill. The catch in the Arctic is zero.

level of peak catches for all individual species because peak catches often exceed a fish population's sustainable catch—the quantity that can be caught indefinitely without harming the prospects of future catches. Furthermore, some of the declines represent major shifts in fish populations from one species to another, either because of natural fluctuations or human-driven

TABLE 2

Fishery Declines of more than 100,000 tons, Peak Year to 1992

Species	Peak Year	Peak Catch	1992 Catch	Decline	Change
		(... million tons ...)			(percent)
Pacific herring	1964	0.7	0.2	0.5	-71
Atlantic herring	1966	4.1	1.5	2.6	-63
Atlantic cod	1968	3.9	1.2	2.7	-69
Southern African pilchard	1968	1.7	0.1	1.6	-94
Haddock	1969	1.0	0.2	0.8	-80
Peruvian anchovy*	1970	13.1	5.5	7.6	-58
Polar cod	1971	0.35	0.02	0.33	-94
Cape hake	1972	1.1	0.2	0.9	-82
Silver hake	1973	0.43	0.05	0.38	-88
Greater yellow croaker	1974	0.20	0.04	0.16	-80
Atlantic redfish	1976	0.7	0.3	0.4	-57
Cape horse mackerel	1977	0.7	0.4	0.3	-43
Chub mackerel	1978	3.4	0.9	2.5	-74
Blue whiting	1980	1.1	0.5	0.6	-55
South American pilchard	1985	6.5	3.1	3.4	-52
Alaska pollock	1986	6.8	5.0	1.8	-26
North Pacific hake	1987	0.30	0.06	0.24	-80
Japanese pilchard	1988	5.4	2.5	2.9	-54
TOTALS:		51.48	21.77	29.71	-58

Source: FAO.

*The catch of the Peruvian anchovy hit a low of 94,000 tons in 1984, less than one percent of the 1970 level, before climbing up to the 1992 level.

environmental change. Commercial fish populations could take decades to recover in the case of such shifts. Nonetheless, these declines indicate the considerable potential to be gained from rehabilitating marine fisheries.[28]

In sum, while the marine fish catch has leveled off since 1989, this tells only part of the story. Behind the stagnant total, the catch of some species continues to rise while that of others is falling. The potential for increasing the total catch from new species is limited because fishers have exploited all of the world's major fishing grounds. The only sure way for fishers to increase their yields—potentially pushing up the marine catch by another 20 million tons to around 100 million tons per year—is to allow fish populations to recover by reducing or temporarily eliminating fishing in many fisheries.[29]

As a result of the slowdown in the ocean catch, people have turned to fish farms for increased supplies. Although people have farmed fish for thousands of years, aquaculture's contribution to the world fish supply has been negligible on a global scale until the last four decades. For the last 10 years, however, the fastest growing portion of the world fish supply has come from aquaculture. Today, as a result of increased freshwater aquaculture production, two freshwater species have displaced marine species in the top ten in the world catch (which includes marine and freshwater species). Silver carp and grass carp have taken over the seventh and tenth positions in the world catch, accounting for 1.6 and 1.3 million tons respectively. (See Table 3 and pages 42-45 for further discussion of aquaculture.)[30]

Lost Bounty

Fishing has long been an uncertain endeavor, pitting the skills of fishers against a sea of variables beyond their control. Sedimentary records reveal wide natural fluctuations in some fisheries. Archaeological evidence indicates fishing societies stressed by major changes in fish populations.[31]

Societies have fished the fertile waters off Peru for at least 6,000 years. Early fishers suffered the same periodic and unpredictable declines that plague Peruvian fisheries today: Every three to ten years, El Niño, an unusually warm mass of water, flows into the eastern Pacific and slows or even stops the nutrient-rich currents, causing lean fishing years.[32]

TABLE 3

Top Ten Species by Weight, 1970, 1980, and 1992 (Catch in million tons)

1970		1980		1992	
1. Peruvian anchovy	13.1	1. Alaska pollock	4.0	1. Peruvian anchovy	5.5
2. Atlantic cod	3.1	2. South American pilchard	3.3	2. Alaskan pollock	5.0
3. Alaska pollock	3.1	3. Chub mackerel	2.7	3. Chilean jack mackerel	3.4
4. Atlantic herring	2.3	4. Japanese pilchard	2.6	4. South American pilchard	3.1
5. Chub mackerel	2.0	5. Capelin	2.6	5. Japanese pilchard	2.5
6. Capelin	1.5	6. Atlantic cod	2.2	6. Capelin	2.1
7. Haddock	0.9	7. Chilean jack mackerel	1.3	7. Silver carp*	1.6
8. Cape hake	0.8	8. Blue whiting	1.1	8. Atlantic herring	1.5
9. Atlantic mackerel	0.7	9. European pilchard	0.9	9. Skipjack tuna	1.4
10. Saithe	0.6	10. Atlantic herring	0.9	10. Grass carp*	1.3

Source: FAO.

*Raised on freshwater fish farms; all others are wild marine species.

While modern-day Peruvian fishers must still contend with El Niño, their own fishing capacity has begun to rival the disruptive capabilities of nature. Heavy fishing combined with the 1972-73 El Niño and a general warming in the equatorial Pacific drove down Peruvian anchovy populations.[33]

With fishers bumping up against the limits of the sea around the world, the ancient scenario of community-wrenching declines is being played out more frequently today than ever before. Because a number of human-driven and natural factors influence fish populations, the best scientists can do to explain fish declines is to assign relative causes and track the population trends. Nonetheless, most fishery and ecosystem decimation clearly stems from marine mismanagement—overfishing above all, but also wasteful fishing practices as well as pollution and habitat destruction. FAO's analysts found overfishing in one-third of the fisheries they reviewed; they found some depleted fish populations in all coastal waters around the world.[34]

Fish populations can tolerate only so much exploitation, based on their numbers, reproductive rate, and death rate—all of which are hard to estimate. If fishers take young fish that are growing rapidly, such "growth" overfishing reduces the potential catch. Fishers can also reduce the overall catch and affect reproduction of the population by removing too many adults—a phenomenon called "recruitment" overfishing. If fishers take so many fish that they destabilize the ecosystem to the point that it experiences significant changes in species dominance, this "ecosystem" overfishing can cause long-term declines in the target species. In what is known as "serial" overfishing, fishers shift from species to species as each is depleted.[35]

Slow-growing species with low fertility are especially vulnerable to depletion. The extinction and near extinction of some whales and other marine mammals are instructive examples of the threat to species with extremely low fertility. Because fish have higher fertility than mammals, fishers have apparently never fished any to extinction, but they have depleted some species to commercial extinction—the point that catches are so low that fishing is no longer economical. Orange roughy, a species from the waters around Australia and New Zealand, takes 32 years to reach sexual maturity. One-third of the orange roughy stocks around Tasmania are reported to be depleted from heavy fishing. Many tuna, shark, and other open-ocean species are also slow to mature and highly susceptible to over-fishing.[36]

All fisheries, however, can be overfished. For example, in the decline of the Atlantic cod and haddock fisheries off North America, people have speculated that the seals may be eating large quantities of fish and that environmental changes or disease may be reducing cod and haddock populations. But here again the dominant cause appears to be long-term overfishing, which has reduced the average size of the cod and haddock, as well as their overall numbers. By removing such a large number of these predators, fishers may have also caused a long-term transformation of the North Atlantic ecosystem. Populations of dogfish and skate—types of shark—have boomed and are now filling the niche left by the cod and haddock. Because dogfish and skate prey on young cod and haddock as well, they are reinforcing this ecological shift. Although the ecosystem is still producing fish, the fishers lose out because there is little demand in North America for dogfish and skate, which do not store well. While other factors may influence the declines, the annual removal of millions of tons of fish appears to be the dominant factor.[37]

Experienced fishers knew that the crash in New England fisheries was coming years before federal regulators stepped in to clamp down on overfishing, according to Penelope Cumler, who wrote an oral history of Maine fishers. Fishers could tell that cod and other groundfish stocks were declining because they caught fewer, smaller fish, and they had a harder time finding even those. One former fisher told her that when he couldn't find many fish in a fishing hole that only he was familiar with, he knew that Maine fisheries were in trouble.[38]

As cod, haddock and other target species start to decline, fishers catch more non-target species such as dogfish and skate. High "bycatch"—the catch of non-target species—is another indicator of overfishing. FAO statistics suggest that total bycatch is growing. The haul of unidentified fish—mostly bycatch—

reached 10.2 million tons in 1992. Another 15 to 20 million tons of bycatch—one-fifth to one-quarter of the 1993 marine catch—is not reported at all: Fishers throw these fish back into the oceans, often dead or dying.[39]

Although overfishing is the biggest threat to fisheries over-all, wasteful fishing practices can damage fisheries just as pro-foundly. Trawling fishers, who drag a large sock-like net through the water, can catch large quantities of non-target species. Shrimp trawlers have the highest recorded rate of bycatch because their nets have a very small mesh size to capture the tiny shrimp. In tropical waters, they bring in 80 to 90 percent "trash fish" with each haul. "Trash fish," however, is a misnomer. The discards include species that other fishers would want but will not be able to catch. Shrimp trawlers in the Gulf of Mexico, for example, net and discard red snapper, undermining a fishery that has already plummeted to one-seventh its potential yield. Worldwide, shrimp fishers are estimated to jettison up to 15 million tons of unwanted fish each year, and other fishers are thought to discard at least another 5 million tons.[40]

Fishing can wreak havoc not only on fish, but on entire ecosystems as well. In the heavily fished North Sea, trawlers have altered the fish populations by hauling out large quantities of adult fish and discarding 2 to 4 kilograms of unwanted species for every kilogram of desired fish caught. Besides causing declines in the populations of commercial species, heavy and wasteful fishing practices may have contributed to the disappearance of porpoises and dolphins from the North Sea by reducing their food stock and drowning them in nets. In the Dutch region of the North Sea, researchers have found that every square meter of the seabed is disturbed at least once a year by trawlers drag-ging nylon nets held open by beams and weighed down by chains. In the heavily trafficked areas, trawlers make an average of seven passes annually with nets that weigh up to five tons each. The nets and chains dig into the seabed and kill sea urchins, starfish, worms, crustaceans, and shellfish—severely damaging the ecosystem.[41]

Around the world, fishing has had a significant effect on wildlife. Before driftnetting was banned, driftnetters ensnared some 42 million seabirds, marine mammals and other non-target species in pursuit of tuna and squid, according to observers for the U.S. National Oceanic and Atmospheric Administration. Tuna fishers in the eastern tropical Pacific Ocean achieved international notoriety because they traditionally sought out and encircled dolphin herds with purse-seine nets to catch the tuna schools swimming below. Purse-seine fishers killed more than 400,000 dolphins per year at the peak of this practice in the 1970s. In the Gulf of Mexico, shrimp trawlers caught endangered sea turtles in their nets. All three of these threats have been largely curtailed thanks to grassroots activism.[42]

Although nets tend to be less discriminating than hooks, fishers also catch birds and turtles on baited hooks laid out by "longline" fishing boats. Japanese tuna vessels operating around Antarctica kill an estimated 44,000 albatrosses annually. Spanish longliners catch an estimated 20,000 loggerhead turtles annually, 4,000 of which are thought to die after the fishers return them to the sea with the hook still in their throat.[43]

One local fisher claimed that his catch of scallops dropped from 300 to 400 in a few hours to 40 or 50 a day because of mangrove deforestation.

As fishers remove an ever greater proportion of the biomass from the marine environment, they can harm wildlife by depleting their food supply. In the North Pacific, the steller sea lion population fell from 300,000 in 1960 to 66,000, when in 1990, the United States government listed them as "threatened" under the Endangered Species Act. Heavy fishing of Alaska pollock, one of their food sources, was a major culprit. Dolphin and bird populations in the region have also declined in recent years as their food supply has dwindled.[44]

In addition to overfishing, the marine environment suffers from pollution and habitat destruction. Several million tons of edible marine fish may be lost to such causes annually. Pollution and habitat destruction disproportionately affect fish that spend at least part of their lives in rivers, bays, estuaries, coastal wet-lands, coral reefs, or semi-enclosed seas: These are the water bodies that are most severely degraded from human activities.[45]

Salmon are among the hardest hit, since they migrate and spawn in freshwater rivers—the most heavily altered water bod-ies of all. In fact, salmon populations began to decline with the rise of industrialization. The salmon population in New England's largest river, the Connecticut, nearly collapsed when the Upper Locks and Canal Company built a large dam on the river in 1798. Similarly, heavy industry in nineteenth-century Scotland led to a decline of the salmon catch on the River Tweed—from over 100,000 to under 20,000 fish a year. Salmon runs throughout the industrial world, particularly along the Atlantic, have disappeared or are threatened by dams, sedi-mentation of rivers, and pollution. In the southern part of the former Soviet Union (the countries surrounding the Black, Azov, Caspian, and Aral seas), water diversions for agriculture have eliminated 90 to 98 percent of the sturgeon, salmon, and other commercially valuable species that migrate through major rivers and estuaries.[46]

The manipulation of rivers can also harm coastal fisheries. Dams, for instance, choke off the supply of vital nutrients from inland to the coast. Egypt's Aswan High Dam, completed in 1965, interrupted the Nile's rich sediment flow into the Mediterranean. The year after its completion, phytoplankton concentrations in the Nile Delta fell by 90 percent, and the sar-dine catch dropped precipitously—from an average of 18,000 tons per year in the early 1960s, to 1,200 tons in 1966, and then 600 tons in 1969. The sardine catch has remained low ever since.[47]

Around the world, development has destroyed an estimated 50 percent of all coastal wetlands—along with many of the creatures that live or spawn in these areas. In Indonesia, the destruction of coastal habitat has eliminated an estimated 60 to 80 percent of the commercially valuable coastal species. The Indonesian government restricted mangrove wetland develop-ment on the south Java coast because of the potential loss of employment and $5.6 million annual income for 2,400 fishers.

In Ecuador, the toppling of mangrove forests for artificial shrimp ponds has undermined the catch of local fishers. One local fisher claimed that his catch of scallops dropped from 300 to 400 in a few hours to 40 or 50 a day because of mangrove deforestation. Mangrove wetlands are vital habitat for numerous fishery species, including shrimp. With half of the world's mangrove forests destroyed, the world's coastal fishers may have lost on the order of 4.7 million tons of potential annual fish catch, including 1.5 million tons per year of shrimp.[48]

The destruction of coral reefs also undermines local fishing efforts in tropical countries. People have destroyed 5 to 10 percent of the world's coral reefs. Given their high productivity, the loss of these reefs may translate into a loss of fish on the order of 250,000 to 500,000 tons per year.[49]

Bays, estuaries, and semi-enclosed seas around the world are at risk from pollution and habitat destruction. In China, pollution's toll on the marine catch is estimated at 210,000 tons a year—and is projected to reach 580,000 tons a year by the year 2000. In the Chesapeake Bay fishery in the eastern United States, the catch has plummeted from pollution and habitat destruction as well as from overfishing. Nutrient pollution from soil erosion and agricultural fertilizers has had particularly severe effects. An excess of nutrients can cause algal blooms that smother other aquatic life. Pollution may also have rendered oysters more vulnerable to the diseases which have overtaken their populations in recent years. Once known as a giant protein factory, the Chesapeake has seen its catch of hickory shad decline 96 percent, alewife and blueback herring 92 percent, striped bass 70 percent, American shad 66 percent, and oysters 96 percent from their historic peaks. When Europeans first came to the United States, oysters in the Chesapeake could filter the equivalent volume of the Bay in two weeks. Now, because so few remain, they take more than a year.[50]

Pollution of coastal waters, however, can take even more insidious forms. The introduction of non-native species is potentially more harmful to the ecosystem than toxic chemicals, according to John Cairns, Jr. and James R. Pratt, co-directors of Virginia Polytechnic Institute's University Center for

Environmental Studies, because these alien species have the potential to migrate between ecosystems and proliferate. In the Black Sea and the adjacent Azov Sea, industrial discharge and agricultural runoff threaten the health of the entire ecosystem— but the 1982 introduction of a jellyfish-like ctenophore, *Mnemiopis leidyi*, has apparently caused even greater destruction. The ctenophore, likely transported to the Black Sea from the Americas in the ballast water of a cargo ship, voraciously devours zooplankton, small crustaceans, and the eggs and larvae of fish—even beyond its own capacity to digest. At times an estimated 95 percent of Black Sea biomass consists of this gelatinous ctenophore. The Azov Sea fishery, which once yielded 200,000 tons of fish per year, is now closed.[51]

Likewise, the invasion of the adjacent Mediterranean Sea by Red Sea jellyfish (*Rhopilema nomadica*) through the Suez Canal has contributed to declines in the Mediterranean fish catch. Other introduced species that have harmed fisheries include the infamous Zebra mussel, ruffie fish, the Asian clam, the Japanese oyster, and marine plants that have altered coastal ecosystems. With several thousand seafaring ships carrying ballast water at any one time, a minimum of several thousand ballast-borne species may be on their way across the oceans on any given day.[52]

In addition, pollution reduces the supply of fish and other marine creatures, not by killing them outright, but by making them toxic. Levels of dioxin in Baltic herring are so high that people can ingest in a single serving what the government considers the maximum safe level of dioxin for one week. The National Food Administration in Sweden recommends not eating Baltic herring or salmon more than twice a week and calls on pregnant women, women who are breast feeding, and young children to forego fish altogether.[53]

Recent research indicates that, in addition to causing cancer, organochlorine compounds such as dioxin can also produce severe hormonal and reproductive problems—including reduced sperm count, babies born with both male and female organs, and other birth defects—both in wildlife and people. Organochlorines disrupt the endocrine system by mimicking natural hormones.

Sperm counts in industrial countries have in fact halved since 1940, while rates of testicular and breast cancer have increased.[54]

Although people can absorb organochlorines through air, water, and other food sources, people who eat fish are especially at risk. Industrial chemicals and pesticides such as atrazine, DDT, dieldrin, chlordane and others eventually flow into bodies of water and contaminate marine life. Fat-soluble chemicals such as organochlorines can accumulate in the fatty tissues of fish higher on the food chain, even if their prey are only slightly contaminated. In northern latitudes, where ocean and atmospheric currents deposit chemicals, sea mammals concentrate high levels of such chemicals. Breast milk from Innuit women who eat these animals on the east coast of the Hudson Bay was found to have four and one-half times the average PCB concentration of the milk from women living in southern Quebec. Other compounds that can accumulate in fish include heavy metals such as mercury, cadmium, and copper, which can cause a range of problems from vomiting and diarrhea to central nervous system and brain damage.[55]

When Europeans first came to the United States, oysters in the Chesapeake could filter the equivalent volume of the Bay in two weeks. Now, because so few remain, they take more than a year.

Sewage and other sources of pathogens can also render seafood unfit to eat. The 1991 cholera epidemic in South America resulted in 300,000 cases and took more than 3,000 lives in Peru alone. Pan-American health officials linked this outbreak to bacteria-laden bilge water from a Chinese freighter. When the ship discharged the water at port, the bacteria contaminated fish and shellfish that people then ate.[56]

Natural toxins from marine organisms, such as the dinoflagellates that cause red tides, can cause gastroenteritis, paralytic shellfish poisoning, and death. In many areas of the world, including the United States, health officials recommend that women of child-bearing age and small children never eat seafood raw. Some 50,000 to 100,000 people a year are hospitalized in the Pacific region because of toxic algal blooms. There are likely an even larger number of unrecorded cases throughout the tropics.[57]

All too often monitoring for contamination is inadequate, and seafood that should not be eaten makes its way to the dinner table. In the United States, where the Food and Drug Administration monitors seafood, risk assessment fails to take into account people who eat a large amount of seafood. The U.S. Environmental Protection Agency's water quality standards and risk assessments are based on consumption rates of one 8-ounce meal per month by a person weighing 150 pounds. This model was designed almost 20 years ago and vastly underestimates average consumption by women, people of certain ethnic groups, and rural and urban poor. Updating the standards to account for higher rates of consumption would lead to stricter permissible chemical standards in water and would better protect the health of fish and the people who eat them. An estimated 20 percent of fish and shellfish consumed in the United States comes from subsistence and recreational fishing (both fresh and marine), which does not receive even a rudimentary level of health-based inspection.[58]

Unlike coastal areas, offshore fisheries do not show direct signs of decline from habitat destruction and pollution. While heavy degradation of coastal areas might be significantly affecting offshore species, scientifically documenting the links is difficult. Scientists have found high levels of PCBs, DDT, chlordane, and other contaminants in dolphins, seals, polar bears and other top predators. Chemicals can build to potentially toxic levels, but scientists have not conclusively proven that chemical pollution is behind recent mass die-offs of marine mammals.[59]

Perhaps the most serious pollution-related threat to offshore fisheries is the buildup of heat-trapping gases in the atmosphere. Atmospheric concentrations of carbon dioxide and other greenhouse gases are increasing as the result of burning fossil fuels and destroying forests. Scientists project that changes in global climate could alter oceanic currents, displacing major fisheries. Polar temperatures are expected to increase more than equatorial ones, diminishing the temperature difference that now drives ocean currents. Scientists suspect that when Europe suffered the so-called Little Ice Age in the 1500s, it was due to a weakening in the Gulf Stream. The collapse of the Peruvian anchovy fishery—partly a result of the slowing of the upwelling waters there—hints at the magnitude of the disruption possible from global climate change. The total productivity of the oceans may not decline, but major dislocations are likely.[60]

Marine fishers, however, do not have to wait for global warming to disrupt their livelihoods. Human action has already caused significant declines in individual marine fisheries. Ironically, the primary problem facing fishers is their own capacity to catch fish. In his 1990 book on marine fishing, anthropologist James McGoodwin from the University of Colorado writes, "Indeed, what is fascinating—and also tragic—about the fishing industry is that it so actively participates in its own annihilation."[61]

Who Will Go Fishing?

With declines in fish populations around the world, the phrase "too many fishers chasing too few fish" has become a cliché. Like many clichés, the statement contains some truth, but also misses a fundamental point. It is not just the number of fishers that counts, but also the size of their nets, the number of their hooks, the girth of their boats—in short, their capacity to fish.

If countries are to control overfishing, a fundamental problem they must confront is the excess capacity of the industry. In facing up to this problem, countries will have to choose which of the three major sectors of the fishing industry to favor: large-scale, industrial fishers; medium-scale; or small-scale, community-based fishers. Each sector has roughly the same capacity to bring in fish. The employment and other social implications, however, are very different.

Reducing the large-scale fishing industry by half would elim-inate some 100,000 jobs. Reducing the medium-scale fishing industry by half would eliminate 500,000 jobs. Reducing the small-scale fishing industry by half would eliminate 7 to 10 million jobs. The trend today is toward industrialization and large-scale fishing vessels: Were this trend to continue, virtual-ly the entire small-scale fishing sector could be wiped out—at a cost of some 14 to 20 million jobs.[62]

Despite the slowdown in the marine catch, the world fishing industry itself geared up greatly in recent decades. Today, world fisheries have on the order of twice the capacity necessary to fish the oceans. Between 1970 and 1990, FAO recorded a doubling in the world fishing fleet, from 585,000 to 1.2 million large boats, and from 13.5 million to 25.5 million gross registered tons. According FAO fisheries analyst to Chris Newton, "We could go back to the 1970 fleet size and we would be no worse off—we'd catch the same number of fish."[63]

Almost invariably, when a country looks closely at its fish-eries, it finds overcapacity. Norway, for instance, estimates that its fishing industry is 60 percent over the capacity necessary to make its annual catch. European Union nations are estimated to have 40 percent overcapacity.[64]

Individual fisheries have shown even greater overcrowding. In the late 1980s, the Nova Scotia dragger (trawler) fishery was esti-mated to have four times the capacity needed to make the year-ly quota for cod and other bottom-feeding fish (groundfish). In the United States, a simulation in 1990 indicated that as few as 13 boats would be sufficient for the East Coast surf clam fishery; at the time there were 10 times that number working the fishery.[65]

How did this overcapacity develop? Many marine fisheries are open to all comers. In its simplest form, open access allows fish-ers to enter a fishery at will. If regulators limit the total catch, they must calculate the potential take of the fishers and adjust the length of the open season accordingly. Fishers then race each other to get the most fish possible. As the number of fishers or their capacity increases, the season gets shorter. In the extreme case of the Alaska halibut fishery, regulators have restricted the season to two or three 24 hour periods per year.[66]

Under open access, fishers continue to enter the fishery well after fish yield and profits begin to fall. As fisheries decline, fishers often buy bigger, faster boats with more advanced equipment and gear. The pressure to overfish, under-report the catch, and even poach can undermine management programs. If the cycle of overfishing and overcapacity continues, profits will decline to the point at which fishers start to go out of business, fewer fishers enter the fishery, and the remaining fishers have no incentive to increase their fishing effort. At this point, if fishing efforts remain constant, the catch of the damaged fishery may stabilize—but at a level below the sustainable potential of the fishery. This point at which the biological and economic factors tenuously balance each other is known as the "bioeconomic equilibrium."

As more and more fishers slide to the brink of financial ruin, pressure on politicians can trigger subsidies that keep overextended fishers in business, maintaining overcapacity. If overfishing becomes too severe, the fishery can collapse, bringing an employment crisis.

In developing countries, open access makes fishing the employer of last resort. People who lose their land or otherwise cannot make a living can always try their luck at fishing. Today, coastal populations are rising faster than total population in many countries, and small-scale fisheries in many parts of the world are being overrun.[67]

FAO estimates that countries have provided on the order of $54 billion annually in subsidies to the fishing industry—encouraging the overexpansion of the industry in the recent decades. The European Union nations, for instance, subsidize their fishing fleets by more than $500 million a year—not including fuel, tariff protection and local government subsidies. Malaysia, having launched a program to modernize its fisheries after independence in 1954, offered subsidies that the World Bank characterized as among the highest in the world. With under 15 million people at the time, the country laid out $30 million for equipment alone between 1977 and 1981.[68]

Because of the importance of fuel in operating costs, fuel subsidies became increasingly common with the oil shocks in the seventies and eighties. Among Taiwanese fishing companies, for instance, fuel accounts for 60 to 70 percent of operating costs, and the Taiwanese government disbursed approximately $130 million in fuel subsidies in 1991. The Soviet Union spent several billion dollars annually on fuel subsidies before its collapse. In the United States, fishers are exempt from paying the $0.20 to $0.22 tax on diesel fuel, which works out to roughly a $250-million annual subsidy.[69]

Besides contributing to overcapacity, government subsidies for the most part favor larger-scale fishers over smaller-scale ones. For example, the Indian state of Kerala pursued a policy of "modernization" in the sixties and seventies that favored commercial fishers over traditional small-scale fishers. Kerala paid for 25 percent of the hull and 50 percent of the engine for commercial fishing vessels and provided low-interest loans for the rest; most of the monies went to more privileged people who knew how to work the government.[70]

But subsidies to small-scale fishers can also be detrimental if they lead to overcrowding and overcapacity. In Kerala, the government reversed its fishery development policy after small-scale fishers started to hold protests and physically threaten commercial fishers. It first eliminated boat subsidies to commercial fishers in 1978, and then started providing small-scale fishers with subsidies for outboard motors, small boats, and modern gear. While the previous subsidies had led to overfishing by commercial fishers, the new policy brought on overfishing by small-scale fishers.[71]

An alternative approach would have been to support the traditional fishers. With its fisheries declining, the Kerala government appointed an expert committee to study the fisheries in 1984. The committee cited overcapacity as the source of the problem and advised emphasizing small-scale, traditional fishing to maximize employment and protect the livelihood of the poorest fishers. The committee recommended reducing the number of trawlers from 2,807 to 1,145, eliminating all 54 boats that use purse seine nets, cutting back on small motorized boats from 6,934 to 2,690, and keeping all 20,000 of the non-motorized craft. If the government had followed its commission's advice, the state of its fisheries might be quite different today.[72]

International development agencies have helped underwrite fishery failures by contributing to the overcapacity of commercial fisheries and undermining traditional fisheries. In a self-evaluation of its own fishery development projects, the World Bank came to the conclusion that the "results have not been satisfactory."[73]

Modernization projects have failed for a number of reasons. One is that development projects aimed at improving the lot of poor fishing communities more often end up serving people who have the resources to take advantage of the new technologies and trade possibilities. Modern equipment can provide the means, and commercial markets the motivation, for depleting fish stocks in ways that are not likely in traditional fisheries. Also, lack of expertise and spare parts can quickly make even outboard motors useless, as can the cost of importing fuel in debt-ridden countries.[74]

In the past, development agencies like the World Bank focused primarily on the purchase of equipment. Traditionally, more than 60 percent of the total aid went toward development of large-scale fisheries, including large vessels, fishing harbors, onshore facilities, as well as technical assistance, marketing and processing capabilities. The major objective of the lenders: to increase production for export and generate foreign exchange. By 1992, a consensus at the World Bank and other development agencies had formed that a new strategy was needed: one that emphasizes management, integration, and public participation.[75]

Modern equipment can provide the means, and commercial markets the motivation, for depleting fish stocks in ways that are not likely in traditional fisheries.

The structure of the fishing industry varies considerably from country to country. Japan, for instance, is the world's top marine fishing country and catches nearly twice as much fish as China: Yet Japan employs only 200,000 fishers compared to

China's 3.8 million. Nonetheless, smaller-scale fishers are almost universally the mainstay of coastal communities because of their numbers.[76]

Around the world, only 200,000 to 300,000 fishers—or about 1 percent of all fishers—work in large-scale fisheries. Another 900,000 to 1,000,000 fishers could be characterized as medium-scale. Of the world's 15 to 21 million fishers, over 90 percent are small-scale fishers, either using traditional equipment or operating small, relatively modern boats. (See Table 4.) Defining these three categories is somewhat arbitrary, but the basic difference is evident from country to country, whether comparing dugout canoes and 20-meter steel trawlers, or the same trawler and a 100-meter factory freezer-trawler.[77]

Although the contribution of all three sectors to the food supply is approximately the same, smaller-scale operations offer a number of important advantages. To catch a given amount of fish, smaller-scale operations tend to employ more people, produce less waste, and require less capital. In addition, smaller-scale fishing supports a greater diversity of coastal communities. Fuel consumption, however, is high on average for small-scale fishers and has been increasing particularly among traditional small-scale fishers, who are buying outboard motors in large numbers. Several million fishers still use non-motorized boats.

On average, small-scale fishers make considerably less money than their more mechanized counterparts. FAO estimates that the crew on the largest boats earn about $15,000 per person per year, while small-scale fishers may garner less than $500 per person per year. As an example of the discrepancies, in Newfoundland fisheries small-scale fishers formed the majority, yet they brought in only 35 percent of the total catch—worth about $8,590 per fisher in 1982. In Asia, traditional fishers are generally poor, despite accounting for one to five percent of national incomes and catching one-third of the fish in the region. An estimated 98 percent of traditional fishers in India fall below the poverty line. Most fishers must also work other jobs.[78]

From a strictly economic perspective, remaining a small-scale fisher may appear to be irrational and inefficient behavior.

TABLE 4

Comparisons Among Fishers by Scale of Operation

Comparison	Large-Scale	Medium-Scale	Small-Scale
Number of fishers employed (million)	0.2 to 0.3	0.9 to 1.0	14 to 20
Fishers employed per US $1 million investment	1-5	5-15	60-3,000
Earnings per fisher (U.S. dollars)	15,000	8,000	500-1,500
Marine fish caught for human consumption (million metric tons)	15-20	15-20	20-30
Marine fish caught for fish meal, fish oil, etc. (million metric tons)	10-20	10-20	Almost none
By catch (million metric tons)	5-10	5-10	Almost none
Annual fuel consumption (million tons)	7.6	12.8	26.2
Fish per ton fuel (tons of fish)	2.6-3.9	1.6-2.3	0.8-1.1*

Source: Worldwatch Institute, based on FAO and other sources in endnote number 77.

Note: Fishers are categorized according to the FAO boat survey described in Marine Fisheries and the Law of the Sea: A Decade of Change. Here, "large-scale" fishers are defined as those who crew boats over 500 gross registered tons; "medium-scale" refers to 100 to 500 gross registered tons, and "small-scale" means under 100 gross registered tons, including traditional boats and canoes. Estimates of small-scale fishers diverge widely partly because they often make a living from various activities.

*A few million small-scale fishers use non-motorized boats.

But the decision to fish is made within the context of fishing communities and cultures. Furthermore, fishers in remote coastal areas have few lucrative employment options. Where there are alternatives, people fish because they like the freedom, the sea, the lifestyle, the continuity of tradition.

Today, however, even the economic efficiency argument does not always favor large-scale fishing. Whereas in years past only the huge boats of large-scale operators could weather the high seas, technological improvements are enabling smaller boats to venture further and further away from shore. Small, powerful boats, miniaturized electronic equipment, and modern fishing equipment allow fishers to do the job of the larger vessels, even traveling hundreds of kilometers out to sea. They can also respond more rapidly to market fluctuations, making them more profitable. In the Mediterranean and the Persian Gulf, smaller boats are replacing the industrial fishing ships because of these factors.[79]

Nonetheless, after years of rapid expansion in marine fisheries around the world, the current trend in country after country is toward consolidation—encouraging bigger boats and smaller fleets. Taiwan has stopped issuing licenses to boats smaller than 1,000 tons and started a buy-back program for boats more than 15 years old. Malaysia plans on using higher capacity, more modern boats—and in the process expects to cut the number of fishers from 100,000 to 60,000. The European Union has set annual targets for individual nations to reduce the number of fishing vessels. As part of Iceland's plan to improve the economic efficiency of its fisheries, the country has undertaken a plan to reduce its fishing capacity by 40 percent. In Canada, the Ministry of Fisheries imposed a quota system on one Nova Scotia fishery that led to the decline in the number of active boats from 455 in 1990 to less than 170 in 1993. Japan has regulated offshore fishing since the end of World War II, and has operated extensive buy-back programs resulting in the scrapping of thousands of boats. With declining fishing opportunities and profits, the government reduced the number of medium-sized trawlers operating in the north Pacific from 97 to 54 in 1986. The remaining trawlers doubled their average individual catches, increasing their profits sharply.[80]

Although consolidation is arguably necessary in many cases, poorly considered programs can eliminate badly needed sources of employment, concentrating the benefits of the fishery in the hands of more privileged people. In the United States, a consolidation program started in 1990 for the East Coast surf clam fishery did not involve crew members or address employment issues. In the course of two years, the number of boats dropped by 53 percent, and the amount of labor by an estimated one-third. Employment would have fallen more if the boat owners had not already begun rotating crew between boats. At the beginning of the program, three firms controlled 33 percent of the surf clam boats, as well as a number of processing plants, while only 21 percent of the boats were owner-operated. The consolidation effort led to even further integration of the industry.[81]

At the request of the boat owners, regional regulators had put in place a market-based system of tradeable quotas, known as individual transferable quotas, or ITQs. Under the ITQ system, each boat owner received a share in the annual catch, and quota holders could buy, sell, or lease them like property. For boat owners, who did not have to pay for fishing rights that they can now sell, the ITQ system yields a windfall profit. Small operators who were having hard economic times were able to sell out or lease their portion of the quota. For the unemployed crew members, the implications are obvious. The unexpected results were that a leaner and presumably more efficient fishery nevertheless did not lower the price of clams—nor did it raise income for most of the remaining crew, despite lengthened working hours.[82]

Although, as the above example shows, the results can be questionable, ITQs are one of the most widely discussed management solutions for overcrowded fisheries. They have a certain appeal because, as transferable fishing rights, market forces can direct the allocation of resources, presumably increasing economic efficiency. For marginal share holders, ITQs have the benefit of allowing them to get out of the fishery with some money. The downside is that such systems allow a small number of individuals or companies to buy control over the fishery. When New Zealand was in the process of instituting an ITQ system, highly capitalized fishing companies expanded their operations beyond what they could sell profitably so that they would account for a higher percentage of the fishery at the time of final allocation. If regulators do not act to prevent such "capital stuffing," ITQs can reward the very fishers who overcapitalized the fishery in the first place, while squeezing out smaller operations.[83]

Limitations on the transferability of an ITQ, such as restricting the portion an individual or company may own, may help to limit consolidation. Under an ITQ system for Alaska's halibut fishery, which is scheduled to begin in 1995, quotas for small boat owners would be allocated in blocks. No single owner could own more than five blocks: In principle these regulations will keep the small boat portion of the quota in the hands of small operators.[84]

But before going the route of consolidation, fishers, communities and regulators should recognize that management systems that promote consolidation also concentrate wealth and can be devastating for coastal communities—particularly if changes come rapidly and without support for developing new jobs. Small-scale fishers are too numerous—and vital to coastal communities—to sacrifice to control overfishing.

Fruits of the Sea: Food & Fairness

While fish supplies stagnate, world population continues to grow approximately 1.6 percent a year. We are adding the equivalent of the population of Mexico annually. At this rate of population increase, the total world supply of fish (marine, fresh, and aquaculture) would have to rise from today's 100 million tons per year to 120 million tons per year by the year 2010, and then add another 20 million tons to the annual supply by 2025 to maintain today's per capita fish supply.[85]

Although the long-term prospects are limited, marine fisheries could continue to contribute to the growth in the world fish supply for the next 20 to 30 years, if well managed. Rehabilitation of stocks could increase the annual marine catch by 20 million tons. Aquaculture, currently growing by approximately 800,000 tons per year, could make up the difference.[86]

The critical question, however, is: Who would benefit from these increases? Current trends—rising prices, increasing exports from developing to industrial countries, and limits on access to fisheries—have severe implications for low-income people and subsistence cultures who rely on marine fish as a dietary staple. Already the distribution of fish is skewed toward consumers in industrial countries, where average consumption per person is three times the level in developing countries. Marine aquaculture has contributed to this disparity. Unless countries manage their marine fisheries for the purposes of maintaining and improving nutrition, increased fish catches will serve only the affluent.[87]

The nutritional benefits of marine fishing are closely tied to the scale of production. Small-scale maritime operations tend to sell or trade their catch locally, particularly in developing countries and traditional cultures; larger-scale operations mostly supply commercial markets, which sell to the highest bidder. This dichotomy has created two global classes of fish consumers. The one linked with local small-scale fishers consists of people with low incomes or in traditional cultures, for whom fish is an integral part of the diet. The class of consumers linked to commercial markets primarily eat fish as a luxury item or supplement to an already balanced diet, as is the case for most consumers in the industrial world.

Unless countries manage their marine fisheries for the purposes of maintaining and improving nutrition, increased fish catches will serve only the affluent.

The differences between these two classes of fish consumers show up in national statistics, although regional differences are even more striking. In countries such as Sierra Leone and the Philippines, fish makes up 50 percent or more of the national consumption of animal protein, and about 25 percent of total protein consumption. Particularly in coastal areas, people in these countries eat fish to raise their overall protein intake to a healthy level. In industrial countries such as the United States and France, however, where protein consumption is twice the recommended level, people on average could greatly reduce or eliminate their fish consumption without significantly affecting their nutrition.[88]

Historically, fish has been considered the poor person's protein because of its relatively low price with respect to meat.

Over the course of the last two decades, however, fish prices have risen relative to beef, pork and chicken because of the combination of rising demand in industrial countries and tightening world supply. (See Figure 2.) Today fish prices are more in line with meat prices. Except perhaps for specialty items such as lobster, if prices rise much further, people will substitute chicken or other meats for fish, so meat prices will constrain fish prices in the wealthier nations.[89]

Consumers in developing countries, however, face a far more dramatic rise in fish prices as their fishers tie into lucrative markets in industrial countries. In Kerala, India's number one fishing state, prices for shrimp skyrocketed from 240 rupees ($50) per ton to 14,120 rupees ($1,300) per ton between 1961 and 1981 with the rise in commercial fishing for export. Per capita consumption fell from 19 kilograms per person in 1971 to 9 kilograms per person in 1981. Sardine and mackerel prices increased ten-fold. Local consumers were no longer competing on the local market with local prices, but on the international market at international prices.[90]

The incentive to export is, of course, cash. In the last two decades, developing countries have increased their share of the marine catch, and in 1989 surpassed the catch of industrial countries. But they are exporting an increasing percentage of their haul in order to gain foreign exchange to pay off foreign debts and import fuel, food, medicine, and other supplies. Exports of ocean products from developing countries have increased twice as fast as those from the industrial countries. Conversely, developed countries import nearly seven times the amount that developing countries import. The trend continues: The government of Vietnam plans to more than double its fish exports by the year 2000, to between $900 million and $1 billion, potentially reducing the supply for domestic consumers.[91]

Increased participation in commercial markets not only raises prices in developing countries; it can reduce the domestic supply of fish, disrupt traditional cultures and lead to hunger. Anthropologist James McGoodwin has documented a case in Mexico where the government limited a local community's fishing rights in favor of more "efficient" commercial shrimpers in the export market. Traditionally fishers, the community had no

FIGURE 2

Comparison of Fish, Beef, Pork & Chicken Export Prices: Growth, 1975-91

Index (1975=100)

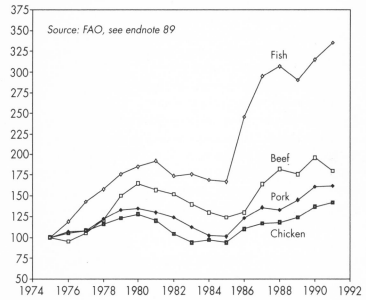

Note: In 1991, the average export price for fish was $278 per ton, versus $284 per ton for beef, $272 per ton for pork, and $163 per ton for chicken.

access to agricultural land. In late summer, when the regulators would close the fishing grounds, the local people went hungry, violent crime increased, children ate dirt or sand to alleviate hunger pangs, and death rates appeared to rise for old people and infants.[92]

When governments subvert traditional, local control of fishing grounds, severe ecological as well as social consequences can ensue. In the Federated States of Micronesia, periods of rule by Europeans and Japanese initiated these changes. The central government subverted local authority to manage fishing grounds, and tie-ins to commercial markets led to the over-exploitation of coral reef fisheries for export. The constant presence of foreign culture has continued to undermine the traditional conservation practices of outlying communities.[93]

Another portion of the catch from the waters of developing countries never touches the shores off which it is caught.

Instead, the foreign fleets pay to fish and ship the catch home. The European Union, for example, pays $200 million annually in access fees, half of it to African nations. Approximately half of the catch off western Africa and one-third of the catch off southwestern Africa is taken by foreign fleets, which return home with most of the fish. In the southwest Pacific, foreign fleets take about one-third, and in the eastern central, the portion is about one-fifth.[94]

Besides catching fish that might otherwise go to local fishers, foreign fleets can also undercut the local marketplace. Commercial vessels sometimes give away fish to gain good will or docking and fishing privileges. Eastern European and other nations give significant amounts of fish to Nigeria and Côte d'Ivoire. Although these giveaways can supplement the domestic food supply, army and government personnel have preferential access. Giveaways can also reduce the market for local fishers, weakening the link between the people who catch and the people who eat the fish, ultimately undermining the economic viability of coastal communities.[95]

The economic power of industrial countries has helped to create this imbalance. The United States, Canada, Europe, and Japan import 84 percent of world fishery imports, by value. Their wealth allows them to continue to gain access to fish by outbidding poorer domestic consumers.[96]

With 123 million well-off consumers who get half their animal protein from fish, Japan is the biggest force in the international market. Still the world's top marine fishing nation, Japan is also the world's number one importer of seafood. With increasing restrictions on where it can fish, the nation has reduced its own catch in recent years while increasing imports. Japanese companies are gaining greater control over processing and marketing, while leaving harvesting to others. In particular, Japan is working to develop links with developing countries that can supply it with seafood. The Japanese semi-public Overseas Fishery Cooperation Foundation (OFCF) coordinates attempts by fishing and trading companies to set up commercial links with developing countries. OFCF offers cheap loans, and the trading companies offer technical assistance.[97]

Massive change—slowing or reversing the flow of fish from developing to industrial countries—could make up for the potential shortfalls in fish supply for the neediest customers. If consumption of fish in industrial countries dropped by 50 percent—which would still allow a generous supply of fish for healthy diets—the supply of fish in developing countries could increase by almost 50 percent. If this transition occurred over the course of the next 15 years, developing countries could maintain their current per capita supply of fish until the year 2010 without other increases in fish supply.[98]

Distribution issues extend beyond the portion of the world catch that goes to people. Approximately one-third of the marine fish catch goes to other uses—primarily animal feed for pets, livestock, and pond-raised fish. The world's number one producer of animal feed is Peru, which annually converts nearly all 6 million tons of its anchovies, jack mackerel, and pilchard to 1.3 million tons of fish meal.[99]

Fishers sell their catch for fish meal when they can catch large quantities easily, but the fish are too small, oily, bony, or otherwise undesirable to command a viable price on the consumer market. This portion of the marine fish catch now benefits primarily the people of more privileged communities, who eat feed-fed livestock or fish. Some also goes into pet food.

But "undesirable" species do indeed feed people and could feed more. In Chile, for instance, despite large gains in jack mackerel production in the past decade, domestic consumption of fish has fallen by half because the fish meal market for export is more lucrative than selling to poor people. Ninety percent of jack mackerel production goes for animal feed.[100]

Yet John F. Kearney, a fisheries sociologist based in New Brunswick, Canada, reports that boys from the neighborhood near a fish processing plant in Talcahuano, Chile, jump into the backs of the moving dump trucks that carry jack mackerel the single block from the wharf to the processing plant. The boys kick as many fish as they can out of the truck before it reaches the plant, while others gather the fish from the pavement to take home or sell on street corners.[101]

For perspective: If the portion of the world catch that now goes for animal feed were offered for human consumption, the transfer would increase the world food fish supply by 40 percent. Such a move would maintain today's world average supply of 13 kilograms per person until the year 2017, without having to increase the supply of fish from other sources. If this fish went solely to consumers in developing countries, it would maintain today's per capita consumption level until the year 2030. Of course, countries would have to exercise care not to disrupt local cultures and markets if such a redistribution effort were ever pursued.[102]

The momentum in marine fisheries, however, is moving in the wrong direction for poorer consumers. Not only have prices risen, but in the past decade the largest increases in supply have come from either low-value species used primarily for animal feed, or high-priced species such as tuna and squid. Neither of these extremes benefit low-income consumers. (See Table 5.) If ever the giant factory trawlers of the socialist countries supplied fish regardless of profit, they are disappearing. Even China has begun to open its giant fish-producing industry to market forces, making its coastal fish farms the number one supplier of shellfish—primarily shrimp—on the international market.[103]

Because of the limits of marine fisheries, aquaculture is gaining attention as an alternative source of fish and other marine products. In general, however, aquaculture serves as a distraction from facing the limits of marine fisheries. Policy makers may be tempted to assume that we can make up for mistreating the oceans and small-scale fishers by farming fish. As mentioned earlier, farmed fish have been the most rapidly expanding portion of the world fish supply in the last 10 years. But their contribution to the welfare and nutrition of coastal people who have traditionally relied on marine fisheries has been minimal.[104]

For instance, the marine aquaculture industry has made extraordinary efforts to increase the supply of such high value species as shrimp. Despite the fact that the wild catch has stagnated around the world, the shrimp supply has continued to grow because of aquaculture. The shrimp farming industry was producing about 500,000 tons per year by the end of the 1980s,

TABLE 5

Top 10 Increases in the Marine Catch, 1970-1989

Species	Increase	Value
	(million tons)	(1989$/ton)
Japanese pilchard	4.7	203
South American pilchard	4.3	90
Peruvian anchovy	3.7	90
Chilean jack mackerel	3.4	90
Alaska pollock	1.7	331
Sandeel	0.8	90
Skipjack tuna	0.7	1,700*
European pilchard	0.6	150**
Yellowfin tuna	0.5	1,700*
Scad	0.4	n/a

Source: FAO.

Note: Atlantic cod, an average-value table fish, is worth about $1,000 per ton.
* Average price for tuna. **Average price for pilchard.

about one-quarter of the total shrimp supply. Similarly, salmon aquaculture was producing about 250,000 tons per year of this high-value species by the end of the 1980s: again, about one-quarter of total production.[105]

A rapidly growing industry, saltwater aquaculture has largely fueled exports. Expanding marine fish farming will do little to meet the needs of the people who are poised to lose as the wild marine supply tightens. An increasingly common practice, in fact, is to catch marine fish and use them as feed for farmed fish. If a person can't outbid a fish raised in a pond for a fish caught in the sea, how can that person then hope to afford the raised fish? Some fishers who sell feed go so far as to use fine-mesh nets to make a clean sweep—biomass fishing. Everything caught is ground up and fed to farmed fish, thus reducing the supply of food fish for local fishers.

Theoretically, aquaculture could employ fishers who lose their jobs, but fish farming involves a quite different set of skills.

Furthermore, the logical alternative for coastal fishing communities is marine aquaculture, which is predominantly tied into the commercial sector. On Java, intense competition in the coastal aquaculture sector fueled consolidation. Many small-scale fish farmers had to sell out as input prices rose, land prices rose, and large-scale fish farms started to underbid them. As ownership became more concentrated and the number of absentee owners increased, fewer workers were employed to raise the same quantity of fish.[106]

Marine aquaculture is also a major cause of coastal habitat destruction, which undermines marine fisheries. Worldwide, one of the major reasons people cut down mangrove forests—half of which have been destroyed worldwide—is to make artificial shrimp ponds. But, as discussed above, coastal wetlands are essential nurseries for wild fisheries, and this destruction directly undermines marine fishing. In Honduras, tension between shrimp fishers and shrimp farmers has led people on both sides to arm themselves; some believe that a conservation-minded fisher was murdered by vigilantes hired by shrimp farmers. Similar conflicts over shrimp farming have flared up around the world.[107]

As well as coastal habitat destruction, shrimp farming and other forms of marine aquaculture have contributed to coastal water pollution, the introduction of alien species and new diseases, and the loss of genetic diversity in wild populations.[108]

For the purpose of feeding needy people while protecting the environment, freshwater aquaculture holds more promise than marine farming. Freshwater fish farms produce less expensive species such as carp and tilapia, which lower-income people are more likely to be able to afford to buy. Carp production in particular has soared in the past two decades. Furthermore, fish farmers have developed farming systems that integrate fish ponds with crop production, so that waste from the ponds fertilizes crops instead of causing pollution. In the Mampong Valley in eastern Ghana—an area characterized by depleted soils, hillside or terraced agriculture, and seasonal rainfall—small farms have benefitted from building fishponds on their marginal and unused land areas. In Vietnam's Mekong Delta, freshwater prawn aquaculture has flourished in integrated systems on farms. Farmers have even combined vegetable and rice crops with marine shrimp and freshwater fish species. These low-tech systems typically consist of agricultural land divided into a series of ditches and levees connected to the main source of water.[109] Fish farming can also aid wastewater purification or recycling. In Bangladesh, farmers grow freshwater fish in ponds and feed them with cultivated duckweed—a common aquatic plant with high nutritional content—instead of soy meal. The duckweed simultaneously processes dissolved nutrients and acts as a natural wastewater treatment system. The pond water can then be used to irrigate cropland and grow rice.[110]

While fish farming has great potential for maintaining fish supplies for inland markets and privileged consumers, the people who currently depend on marine fisheries are not poised to reap many benefits. The greatest hope for continuing to meet the needs of coastal peoples lies in the small-scale fishers who currently serve them. Therefore, rehabilitating marine fisheries and maintaining access for small-scale fishers is not only a matter of employment and community support, but also a question of nutrition for many of the world's poorest fish consumers. For these people in particular, the tightening in world fisheries is increasingly threatening, and without preventive action, marine fisheries will cease to serve much of their current vital nutritional role.

Fencing the Oceans

As one of the great global commons, the oceans fall within the realm of international governance. One of the fundamental challenges to managing marine fisheries is determining who has jurisdiction where. The countries of the world have hammered out some important conclusions, but fundamental issues remain ambiguous.

Only in modern times has managing marine fisheries become a contentious international issue. For centuries, local fishers controlled the most heavily fished waters along the coast; the high seas remained open to all. In 1609, Dutch legal scholar

Hugo Grotius articulated this common law in a treatise, *Mare Liberum*, that was generally accepted by the international community. Grotius asserted that outside coastal territorial waters—defined by what a country could defend—all people should be able to act freely, partly since the ocean and its resources were greater than people could alter. Territorial waters became commonly defined as 3 nautical miles (5.6 kilometers).[111]

Competition over high seas fishing grounds started to heat up by the turn of the 19th century. With the boom in marine fishing after World War II, fishers began to engage in sometimes fierce competition over fishing grounds beyond territorial waters. In the seventies, as conflicts mounted, the world community agreed that freedom on the high seas—at least in the realm of fishing—is an anachronism. As part of the Third United Nations Conference on the Law of the Sea, the governments of the world agreed to establish a zone no more than 200 nautical miles wide, within which the coastal country has exclusive rights to natural resources. Known as exclusive economic zones, or EEZs, these areas include the most productive fishing grounds in the oceans. By 1976, more than 60 countries had laid claim to the waters 200 nautical miles from their shores. Today, more than 122 nations have claimed EEZs. (See Table 6.)[112]

The EEZ system is a compromise. It removes the richest portions of the oceans from the global commons and confers sovereignty over them to individual coastal nations. These countries are arguably in the best position to manage the adjacent fisheries, but land-locked and other nations lose access to what had been open territory. In addition, EEZs give both industrial nations with island protectorates, and small, independent island nations greater control over the oceans. And while resolving some conflicts, the EEZ system exacerbates others.

The United States has the world's largest EEZ partly because of its long, unobstructed coastline, and partly because of the numerous islands under its control. France has the second largest EEZ in the world, mainly due to its island protectorates.[113]

EEZs also give new international clout to small island nations. Pacific island nations do not have the highly productive fishing grounds associated with continental shelves, but their com-

TABLE 6

Top 15 of 122 Exclusive Economic Zones (EEZs)

Country	Marine Area (sq. nm)	Portion of Total EEZ (percent)
USA	3,107,000	10
France	2,100,000	7
Australia	1,854,000	6
New Zealand	1,792,800	6
Indonesia	1,577,300	5
USSR (1991)	1,309,500	4
Japan	1,126,000	4
Brazil	924,000	3
Canada	857,000	3
Mexico	831,500	3
Kiribati	770,000	2
Chile	667,300	2
Norway	590,500	2
India	587,600	2
Philippines	551,400	2
Subtotal (Top 15 EEZ Areas)	18,645,900	59
Total (122 EEZ Areas)	31,464,400	100

*All EEZ areas together total less than 10 percent of the oceans' surface, but contain virtually all the prime fishing grounds.

Note: As of February 1992, 122 coastal states had claims for EEZs extending beyond a 12 nautical mile (nm) territorial sea.

Source: Judith Fenwick, *International Profiles on Marine Scientific Research* (Woods Hole, Mass.: Woods Hole Oceanographic Institution, 1992.)

bined marine jurisdictions give them significant control over the Pacific Ocean. The island nation of Kiribati, for instance, has the world's eleventh largest EEZ—1,126 times its land area. Because of their extensive legal jurisdiction, these island nations were a major force behind the international ban on driftnetting, the wasteful fishing practice that entangles birds, marine mammals and other non-target species in enormous nets. The regional organization, the South Pacific Forum, advanced the issue by banning the use, possession, and transit of driftnets longer than 2.5 kilometers in the waters and territory of member Pacific island nations. The Forum then helped bring the issue before the United Nations, which enacted an international moratorium on driftnets that went into effect on December 31, 1992.[114]

The EEZ system added fuel to the ongoing dispute between Argentina and Great Britain over the Falkland, South Georgia, and South Sandwich islands. The contested islands give Britain control of a portion of the rich squid fishing grounds on the Patagonia continental shelf in the southwestern Atlantic. Licenses to Japanese, Taiwanese, and Korean fishing vessels around the Falkland Islands bring in $45 million per year, more than any other source of revenue.[115]

The Law of the Sea officially goes into force in November 1994. But it remains vague on the matter of fishing the high seas outside the EEZ limits. For example, small portions of Canada's Grand Banks extend more than 200 nautical miles into the Atlantic. The Canadian government accuses foreign fishers who work these waters of depleting cod stocks. In one case, Canadian enforcement officials boarded and seized a trawler, the Kristina Logos, despite the fact that the boat was 45 kilometers outside Canada's EEZ.[116]

The Kristina Logos had taken advantage of the most significant fisheries loophole in the Law of the Sea: a dearth of enforceable restrictions on fishing on the high seas. Regional fishing agreements cover various high seas fisheries, including the one outside the Grand Banks, but these lack teeth beyond the power of diplomacy. If a country does not wish to comply with restrictions imposed by an agreement, it can refuse to participate, as Norway did with the whaling moratorium; or it can simply quit, as Iceland quit the International Whaling Commission. When the home nation of a vessel enters into a regional agreement that might restrict fishing, the vessel's owners can avoid the restrictions by adopting a flag of convenience—a flag of a non-signatory to the agreement. Such was apparently the case with the Kristina Logos, whose Portuguese captain and crew were flying a Panamanian flag.

The most dramatic failure of high seas diplomacy is the case of the beleaguered bluefin tuna. Bluefin tuna are the world's most valuable fish, fetching up to $260 per kilogram in restaurants in Japan. In principle, the International Commission for the Conservation of Atlantic Tuna (ICCAT), an organization of nations whose fishers hunt tuna in the Atlantic Ocean, manages bluefin tuna fishing. But despite the oversight of ICCAT, bluefin tuna populations have dropped precipitously. The population that spawns in the Gulf of Mexico has dropped by 90 percent since 1975, and the one that spawns in the Mediterranean has declined by half. The ICCAT governing board, which represents fishing interests, stalled on lowering the bluefin quotas until 1993, after Sweden proposed listing bluefin tuna as endangered and threatened under the Convention on International Trade in Endangered Species (CITES). Even then, ICCAT chose a phased cut of 50 percent by 1995. CITES listing would have banned international trade of the western bluefins and placed the trade of eastern bluefins under close monitoring.[117]

Regional fishing agreements cover various high seas fisheries, including the one outside the Grand Banks, but these lack teeth beyond the power of diplomacy.

The prospects for bluefin tuna recovery remain dim. ICCAT members complain that up to 80 percent of the catch may be taken by non-ICCAT fishing vessels flying flags of convenience— some 500 of which are thought to be hunting bluefins in the Atlantic. If so, the ICCAT action is unlikely to make much difference. Trade restrictions under CITES would be more effective since fishers outside the ICCAT system would not be able to legally sell their catch. But the United States, Canada, Japan and other ICCAT member nations have blocked such restrictions.

Japan—the major buyer of bluefins—proposed banning imports from non-ICCAT nations under the ICCAT system, but met with resistance from the European Union, which includes several non-ICCAT nations.[118]

The basic tenets of freedom of the high seas, coupled with a tightening situation in world fisheries, have heightened international tensions around the world. (See Table 7.) A major area of conflict involves straddling stocks—fishing grounds that extend beyond the 200 nautical mile limit. Straddling stocks are particularly common near continental shelves that thrust past 200 nautical miles, such as Canada's Grand Banks and the Patagonia Shelf. Conflict also erupts over highly migratory stocks, particularly tuna, which range over large areas of the ocean. In both cases, a country may claim jurisdiction only if it can prove that the fish originate in its waters. Salmon, for example, may be claimed since they spawn inland in rivers, but other fish remain fair game to all.[119]

Conflicts over high seas stocks precipitated international negotiations at the United Nations beginning in July 1993. To date the negotiators have been unable to reach agreement. Two camps have formed—one primarily of countries with major fishing grounds who want a binding international treaty, and one mainly of nations with long-range fishing fleets, who want only a guideline.[120]

A regime to regulate fishing on the high seas would mean the end of the freedom of the seas doctrine, at least with regard to fishing. Just as the western frontier of the United States disappeared behind fences, the oceans will inevitably be divided up by diplomacy. A regime for regulating the high seas could take various forms: Rights to the oceans could be distributed among the nations of the world, or an international body could govern all fishing, or—the most likely resolution—all fishing vessels could be required to fly the flag of a nation that has signed regional agreements. The sticking point is membership: Could any country join? Long-time fishing nations complain that new entrants would divide the oceans into unrealistically small shares; developing countries fear being shut out because they have not yet built the capacity to fish the open oceans.

TABLE 7

International Conflicts over Fishing Grounds

Region	Conflict
Northwest Atlantic	Canada's Grand Banks extend beyond the 200-nautical-mile EEZ. Ships with flags of convenience fish depleted cod stocks outside the EEZ. (See text.)
Northeast Atlantic	Various disputes over fisheries between Greenland, Norway, Iceland, Great Britain, and other European Union nations. Also arguments within European Union over trade and fishery policy.
Atlantic Ocean	High seas competition for dwindling bluefin and other tuna stocks. Members of the International Commission for the Conservation of Atlantic Tunas in conflict with non-members. (See text.)
Southwest Atlantic	Argentina and Britain contest the Patagonia Shelf, which extends from Argentina past the Falkland Islands. (See text.) European and Asian nations also fish the shelf beyond the EEZs.
West African Coast	European Union, Eastern European, and former Soviet fleets have heavily fished the rich continental shelf, causing with the coastal nations.
Namibian Coast	Spanish and other foreign fleets overfished the fertile Benguela current. After gaining independence in 1990, Namibia asserted its authority and cut quotas drastically.
Southeast Asia	China, Taiwan, Vietnam, Malaysia, the Philippines, and Brunei assert conflicting claims over the South China Sea.
North Pacific	Foreign fleets have heavily fished the "Donut Hole" in the Bering Sea between the EEZs of Russia and the United States. The United States, Russia, China, Japan, Korea, and Poland have tentatively agreed to halt fishing in the area.
North Pacific	The "Peanut Hole" in Russia's EEZ in the Sea of Okhotsk has been heavily fished by Korean and Chinese fishers, as well as vessels flying flags of conveience, particularly Panamanian.
Global	United Nations moratorium in effect on driftnets longer than 2.5 kilometers. Most nations have reduced driftnetting, but a few reportedly continue and Italy is using 50 kilometer nets in the Mediterranean.

Source: Worldwatch based on sources in endnote number 119.

Equity dictates the participation of all nations—either by being allowed to fish, or by receiving compensation if they choose not to. Successful precedents exist for such agreements. One of the first and most fruitful international agreements on the capture of marine life was the 1911 Convention for the Preservation and Protection of Fur Seals. Russia, Japan, the United States and Canada (represented by Great Britain) negotiated the agreement after uncontrolled hunting caused the seal population to plummet. The four signatory countries agreed to allow regulated hunting of the North Pacific fur seal only on specific breeding islands. In return for giving up the right to hunt seals at sea, Japan and Canada each received 15 percent of the skins as compensation from the United States and Russia, which managed the seal hunt. Under this arrangement, the seal population soared from 125,000 in 1911 to 2,300,000 in 1941.[121]

An alternative would involve expanding the EEZ concept. Here again precedent exists: The Law of the Sea specifies that even though the EEZ ends at 200 nautical miles, nations own exclusive rights to the mineral resources of the continental shelf out to 350 nautical miles. Extending the EEZ that far would cover a number of areas of existing conflict.[122]

While the debate continues, FAO is working to improve international cooperation through other fora. The U.N. agency adopted a set of guidelines on reflagging at its annual meeting in November 1993. FAO also convened a multinational panel to devise a code of conduct on responsible fishing, a draft of which was released in February 1994. Both of these documents are nonbinding guidelines, since FAO has no power over individual nations. But its expertise in world fisheries does give the body considerable credibility and stature. Further revision of the centuries-old concept of freedom of the seas may be slow in coming, despite FAO's efforts, but the combination of the obvious limits of the sea and the number of vessels trying to get a portion of the catch will keep pressure on the governments of the world until they act.[123]

Fishing Within The Limits

Two decades after the nations of the world convened for the Law of the Sea talks, coastal countries control the prime fishing grounds—but the number of fish populations that have collapsed or are at risk is higher than ever. Coastal countries have replaced overfishing by foreign fleets with overfishing by their own. For many fishers and their fisheries, the crisis point that nations worked to avert 20 years ago has arrived.

Worldwide, effective fisheries management would not only save jobs: It would also save tax-payers tens of billions of dollars per year. Governments could potentially save some $54 billion per year by eliminating subsidies, and earn another $25 billion per year in rents, with a net budgetary benefit of more than the current gross value of the entire marine catch. Meanwhile, if stocks are allowed to recover, FAO estimates that fishers could increase their annual catch by as much as 20 million tons—worth about $16 billion at today's prices. Although this theoretical exercise does not take into account the broader adjustments that societies will have to undergo to redirect former fishers into new occupations, it conveys the magnitude of economic mismanagement behind the ecological mismanagement of the oceans.[124]

To make the transition to healthy fisheries, governments and fishers will have to move beyond the current state of political deadlock. In virtually all cases, a combination of government oversight and community-based management promises the best solution. Local administration backed by governmental authority would allow the parties most involved the clout to regulate fisheries for the good of the local economy and the ecology that supports it.

To put fishing on a sustainable path will mean at least a slowdown in major fisheries while fish populations recover. Programs to help ease the transition are vital. A few are already being assembled, including a $30-million package for New England fishers and their communities. Twelve million dollars of this is earmarked to help individual fishers move into other fisheries and other industries. If, however, such programs simply encourage fishers to move to other fishing grounds or species that are not yet overexploited, then the vicious cycle of reaching out to new fisheries and depleting them may continue.[125]

The basic tenets of fishery management were developed over thousands of years by traditional cultures who relied on fish for food. In many traditional Pacific island and Southeast Asian coastal cultures, for instance, limited access was an integral part of maintaining the productivity of coral reef fisheries, which are highly vulnerable to overfishing. Typically, a master fisher would regulate fishing with closed seasons, restricted areas, size limits, species restrictions, quotas, and equipment regulations—all of which prevented overfishing and allowed reef species to repopulate. Breaking the taboos against overfishing could lead to expulsion from the community or death.[126]

Most of these restrictions still form the basis of fisheries management. The indispensable element all too often lacking today is local, community-based control. Fishers readily subvert management systems that do not involve them. In addition to outright illegal methods of enlarging their catch, fishers bend the rules by increasing their fishing capacity with bigger boats, nets, and so on. They get around limits on boat length by buying wider boats. In fisheries with restricted boat length, the boats become almost as wide as they are long.

Short of Orwellian monitoring, centralized management fails. Examples of successful management involve a high level of fisher and community involvement. In Maine, the lobster fishers developed their own effective system of limited access without any government involvement. Each local harbor has its own territory, which is further subdivided among the lobster fishers. The system's success is based on the tight-knit communities. The rules are taken so seriously that violence occasionally flares when fishers break them.[127]

Japan manages its coastal fisheries under a two-tiered system with roots in village customary law of the feudal era. During the Edo Period, from 1603 to 1867, the nation developed detailed fishing regulations and institutionalized its system of local sea tenure. For example, seaweed harvesting was banned during spawning season to protect the fish eggs attached to seaweed, gill nets for bottom species were outlawed, and night fishing with torches was limited during the mid-1800s. As the government of Japan became increasingly centralized, local communities nevertheless continued to control coastal fisheries. Japan passed the national Fisheries Law in 1901 to formalize the existing system of control and access through the Fishing Cooperative Associations (FCAs).[128]

Essentially, the FCA owns the local fishing grounds; all members have a share just as a stockholder owns a share of a company. The FCAs form the link between the government and local fishers. FCAs organize all coastal fishers and enforce control over the fisheries.

Because the Japanese system is grounded in the feudal past, it still smacks of elitism. People without a family connection have a hard time entering fishing. But long-time Japanese fishers are relatively well-off, in sharp contrast with fishers in many parts of the world. And the basic two-tiered structure of fisheries management—stable for hundreds of years—forms a model promising for other countries as well. Higher levels of government set guidelines, but local people work out the details.[129]

To the extent that a fishery management system gives fishers a strong sense of ownership, fishers will have more of an incentive to steward the fishery for the long term.

Similarly, small-scale fishers in the Maluku Islands in Indonesia have modified their traditional management system, known as *sasi*, to adapt to changes brought by interaction with commercial markets. *Sasi* combines management and spiritual practices to maximize the catch. Beginning in the sixties, the fishers began gathering *trochus*, a reef mollusk, to export the shells to Italy, Japan, and other Asian markets for buttons and pigments. Overharvesting in the eighties led to declines in *trochus* catches, which the fishers blamed on failing to please ancestral and environmental spirits. The government stepped in and banned the harvest in 1990. Over time, the local fishers modified *sasi* to take into account scientific notions of environmental dynamics—and *trochus* gathering began again on a sustainable basis.[130]

Enforcement is critical to the success of any management system. To the extent that a fishery management system gives fishers a strong sense of ownership, fishers will have more of an incentive to steward the fishery for the long term. For example, in the United States individual fishers often lease their own shellfish beds. Therefore close oversight is unnecessary because it is the individual fisher who stands to lose if he or she overexploits it. Likewise, a community-based fishery can be self-policing if the community is tight-knit and understands the ecology of the fishery.

Market-based systems that give individual fishers a share in production (such as ITQs, discussed above) can also give fishers a long-term interest in the fishery. But if the fishing grounds lie far offshore, the species move around, and there are many fishers, any given fisher is likely to act as if in an open access system. Individual quotas can even make enforcement more difficult because fishers can fish when they want. With an open-ended season, fishers can get around the system by not reporting catches, so they can exceed their quotas, and by high-grading—discarding lower-value fish and keeping only the more valuable ones to maximize the value of their catch. A defined season can be easier to manage and lead to less abuse because enforcement officials know that fishing in a particular region is only allowed at certain times. Likewise, protection of closed areas, which can serve as undisturbed habitat for reproduction and growth, can be straightforward. On the other hand, clearly defined catch limits for each fisher remove the race for fish, which encourages fishers to overinvest in bigger, faster, higher capacity boats to out-compete each other, and which is potentially dangerous and wasteful.

Ultimately, fisheries are part of the public trust, and governments have a responsibility to maintain them for future generations. Ideal management systems would require little government participation, but there is always a role for the government, whether enforcing the right of a community to bar outsiders and manage its own fisheries, or more active regulation and patrolling.

In the Philippines, for instance, the government grants local communities 25-year contracts to manage the adjacent coastline. With the authority of the government behind them, several communities have restored hundreds of hectares of mangroves, established no-fishing zones, and limited fishing—with resulting increases in the sustainable fish catch. When the communities did not have government backing, commercial fishers encroached on their territory.[131]

Developing countries often lack resources for management and enforcement programs, however. Modern patrol boats can be as hard to maintain as modern fishing boats, and managers may lack expertise. In hopes of improving management, the World Bank fishery program co-sponsored a management symposium with the Peruvian Ministry of Fisheries in June 1992. The meeting brought together fishers, scientists, government officials and environmental activists. As an example of the management cooperation possible, The Gambia in 1990 entered into a joint patrolling agreement with China, under which China helped halve the number of boats fishing for squid in Gambia's waters and greatly reduced poaching.[132]

Developing countries commonly establish a coastal zone where only traditional, small-scale fishers may fish. But these zones are vulnerable to encroachment and fishing further offshore by commercial fishers. In Sierra Leone, the catch of traditional fishers dipped significantly as commercial fishing intensified. Greater control over offshore fishers would directly benefit small-scale fishers.[133]

Once a country decides how to manage its fisheries, the next step is to make sure that financial incentives support the overall strategy. All too often, governments subsidize overcapacity and overfishing even as fish catches decline. FAO estimates annual subsidies to fishing to be on the order of $54 billion annually, more than two-thirds the value of the annual marine catch.[134]

In addition to subsidies, governments lose money they might otherwise make from fisheries, since the depleted fisheries do not yield as much as well-managed fisheries would. In Europe, lost revenue due to overcrowding of fisheries is estimated at $2.5 billion annually. Economic analysis of Iceland's fishing industry indicated that the country could increase its gross national product by four percent by managing fisheries more efficiently.[135]

Further, governments often forego potential royalties for the use of fishing grounds, which are after all a public resource. As in the management of grazing, logging or mining on public lands, fees can be an integral part of limiting exploitation and compensating the public for use of commonly held resources. At a minimum, fisheries should be self-supporting, the fees covering management. If a community decides to take a market-based approach to its fisheries, the public can require fishers to pay for tradeable fishing rights, which would also eliminate windfall profits from giving away rights that can then be sold.

In Australia, rents for the use of fishing grounds have ranged from 11 to 60 percent of the gross value of the catch, with a weighted average of 30 percent. In an innovative tradeable quota system, the Australian government takes a percentage of each fisher's allotment each year, which it then sells to cover management costs and to give new fishers the opportunity to enter the fishery.[136]

Governments can adjust royalties according to social and conservation goals. For example, small-scale fishers could be exempt from payments. Fees for commercial fishers can increase as the stocks become more depleted. Many countries already charge foreign fishing vessels for access to their fishing grounds. Domestic charging is much less common. In the United States, licensing fees fail to cover the cost of fishery management, even for foreign vessels.[137]

Countries may have rich marine resources from which they derive little benefit because of mismanagement, poaching, and corruption. After studying fisheries management in West Africa, Vlad Kaczynski at the University of Washington School·of Marine Affairs concluded that foreign fishing fleets were essentially cheating poor nations. Guinea, for instance, received about 1 percent of the value of the catch by foreign vessels, compared to 25 percent royalties in many industrial countries. This foregone revenue could fund better fisheries management, including enforcement to protect traditional fishing grounds from encroachment. Some of the revenue could go to alternative employment training in coastal communities, redistribution costs, and fisheries research.[138]

In addition to reducing overfishing through improved management, fisheries need to operate within the context of a comprehensive oceans policy. The two key components of such a policy are reducing the environmental effects of fishing, and reducing the effects of broader environmental degradation on fishing. Both parts would greatly benefit from fisher participation.

Bycatch and other waste are the number one issues for reducing the ecological effects of fishing. Because fishing varies widely, the techniques for reducing waste are specific to each fishery. In shrimp trawling, the fishing practice with the highest bycatch rate, fishers have had excellent results with devices that keep out turtles and other larger species. In the United States, shrimp fishers off the northeast coast originally resisted using the Nordmore grate because they thought it would reduce their

With 77 percent of marine pollution coming from land, and half the world population living in the coastal zone, these waters are subject to degradation.

catch. But then they discovered that it actually made shrimp trawling more efficient because fewer unwanted species wound up in the nets.[139]

In Alaska, fishery managers are considering incentives for fishers to devise their own methods for reducing waste. Under a program called Harvest Priority, fishers who can document a lower level of bycatch than the fishery average would receive the right to additional fishing time or quota, depending on how the fishery is managed. Harvest Priority would harness fishers' knowledge and expertise without burdening them with government regulations.[140]

Alaska fishery managers are also considering a system that requires fishers to fully utilize everything they catch. Currently, regulation itself fosters a large proportion of bycatch: Regulations often divide fisheries so that fishers have to throw back non-target species—often dead or dying—even if they are commercially valuable. Factory ships also discard species too big or too small for their automated equipment to handle. Full utilization programs would encourage fishers like the shrimp trawlers to reduce their bycatch. Management would be more complicated under this system, but the potential for reducing waste makes full utilization a goal worth working toward.[141]

General environmental degradation is also a threat to marine fisheries. As we have seen, 90 percent of the marine catch comes from the 10 percent of the oceans nearest the continents. With 77 percent of marine pollution coming from land, and half the world population living in the coastal zone, these waters are subject to degradation. Coastal habitat is particularly important because of its high productivity and the role it plays as the ocean's nursery. Some two-thirds of commercially caught marine fish spend part of their early, most vulnerable stages in coastal habitats. Pollution and habitat destruction affect fisheries directly, and climate change has the potential to dramatically transform marine fisheries in a number of ways, ranging from altered oceanic currents to more severe storms.[142]

It is in fishers' interest to work for environmental protection of the ocean. Subsidies that encourage environmentally destructive practices on land can affect their livelihoods. Water development subsidies, for instance, have led to the damming and diversion of rivers, ruining salmon and other diadromous fish habitat and causing catches to plummet. Subsidies for logging and agriculture can promote pollution of rivers, as well as the estuaries and coastal habitat on which many marine species rely. Subsidies for coastal development can lead to the direct destruction of coastal wetlands.[143]

Typically, fishers have little or no say in what happens to the waters they fish. Economic development typically takes precedence over fishing interests. On the Texas coast of the Gulf of Mexico, shrimp trawlers and oyster harvesters risk greater reductions in catch as the coast continues to attract—and the state continues to permit—more petrochemical and other industrial facilities. With the building of one new plastic manufacturing plant, for instance, the state indefinitely prohibited nearby oyster harvesting while monitoring the effects of the increased discharge of chemical-laden wastewater.[144]

Japan is the exception: Not only do fishing communities control coastal fishing, they also have authority over coastal development. If a company wants to build along the coast, the local fishers have the right to block the plans or demand compensation.[145]

Such power, however, is unheard of elsewhere in the world. Nonetheless, fishing organizations are increasingly recognizing the importance of protecting the marine environment. In the Philippines, Honduras, and elsewhere, small-scale fishers have worked for the conservation of mangrove forests and coral reefs. In Mexico, fishermen in Lazaro Cardenas, one of the country's major ports, are organizing to try to prevent further deterioration of their fishing grounds, which they claim began as the government encouraged industrial development in the area during the seventies. Oil companies, steel factories, and fertilizer plants have dumped untreated toxic wastes into the rivers and coastal waters, destroying estuaries and decreasing their fish catch. The fishers reached the limit of their tolerance when a Norwegian ship carrying 9,000 tons of sulfuric acid spilled acid in the port, was dragged out to sea, then washed up on the beach during Hurricane Calvin. Eight months later, the ship remained beached with an estimated 4,500 tons of sulfuric acid in its hold. Fishers' demands for compensation were denied, and the only information the government issued was a 44-day prohibition on fishing near the ship. In protest, the fishers took over the port. Twenty-four were jailed and three leaders were held; the community received no compensation for the environmental damage.[146]

These issues of fishery management, however, are far from new. People have known for thousands of years how to manage fisheries for maximum sustainable yield. The most difficult choices today are the political ones. Since the existing government and fishing industry policies led to overcapacity and over-exploitation, new perspectives are called for. Such perspectives will largely come from community participation, not just from small-scale fishers—who are typically outside the political process—but also from public interest groups that raise issues of equity, social stability, and environmental protection.

Given the overcapacity that exists today, few countries will be able to avoid losing jobs. But they can orient their policies so that job-losers number 1 or 2 million large- or medium-scale fishers—or 5, 10 or 20 million small-scale fishers and the communities they support. Because of the social consequences, national governments would do well to help keep small-scale fishers in business.

Fishers, regulators, and coastal communities are at a crossroads. If they continue on the current path, marine fisheries will continue to decline, millions of fishers will lose their jobs, and coastal communities and low-income consumers will suffer disproportionately. If instead these groups combine forces to improve fishery management, the oceans can continue to yield fish—and economic and social benefits—for centuries to come.

Notes

1. Colin Nickerson, "'Pirates' Plunder Fisheries," *Boston Sunday Globe*, April 17, 1994; job estimates range from 30,000 to 50,000 fishers, from Mark Clayton, "Hunt for Jobs Intensifies as Fishing Industry Implodes," *Christian Science Monitor*, August 25, 1994.

2. Clyde H. Farnsworth, "Canada Acts to Cut Fishing by Foreigners: Will Seize Boats Outside Its Waters," *New York Times*, May 22, 1994.

3. Maine lobster example from "There's a Catch," *The Economist*, September 18, 1993; Texas example from Sam Howe Verhovek, "Shrimpers Feel at Bay Over Plant Expansion," *New York Times*, June 20, 1993.

4. Regional analysis based on statistical data in the United Nations Food and Agriculture Organization (FAO) fisheries database, FISHSTAT-PC, FAO Fisheries Statistics Division, Rome, 1994.

5. Quote from FAO, *Marine Fisheries and the Law of the Sea: A Decade of Change*, FAO Fisheries Circular No. 853 (Rome: 1993).

6. Canada has lost some 30,000 to 50,000 fishing jobs; New England is likely to lose around 20,000 jobs from Elizabeth Ross, "Hard-Hit New England Fishermen Receive Financial Aid," *Christian Science Monitor*, March 23, 1994; U.S. Pacific salmon fisheries have lost on the order of 60,000 jobs, from Mark Trumbull, "Pacific Northwest Fisheries Shrink, Taking Thousands of Jobs Along," *Christian Science Monitor*, March 28, 1994; and China's leading fishing province, Guangdong, has lost 14,000 jobs, from Fan Zhijie and R.P. Côté, "Population, Development and Marine Pollution in China," *Marine Policy*, May 1991. In addition, European nations have lost at least several thousand fishing jobs, and the disbanding of the Soviet Union has put an uncounted number of fishers out of work, as have the decline of the Black Sea fisheries and the closure of the Azov Sea fisheries; the environmental collapse of the Aral Sea put 60,000 people out of work since the 1960s. Potential for future job loss discussed later.

7. Southeast Asia from Mohd Ibrahim Hj Mohamed, "National Management of Malaysian Fisheries" *Marine Policy*, January 1991; Leslie Crawford, "Chile No Longer has Plenty More Fish in the Sea," *Financial Times*, July 19, 1991; Gylfi Gautur Pétursson and Kristján Skarphéðinsson, "Restructuring the Fishing Industry in Iceland," Ministry of Fisheries in Iceland, Reykjavík, 1992; 200 million from Lennox Hinds, "World Marine Fisheries," *Marine Policy*, September 1992.

8. Availability varies from region to region. A number of species are not readily available fresh, such as halibut from Alaska and rockfish from the Chesapeake. Fresh salmon and catfish caught in the wild are not readily available in Washington, D.C., although farmed species are. Based on informal survey of Washington, D.C. supermarkets, June 24, 1994.

9. FAO, *Marine Fisheries*, op. cit., note 5; marine fish supply from Edmondo Laureti, FAO Fisheries Department, *Fish and Fishery Products: World Apparent Consumption Statistics Based on Food Balance Sheets (1961-1990)* (Rome: FAO, November 1992). At approximately 70 million tons per year and 52 million tons

per year, respectively, pork and beef production are second and third to marine fish production of 80 million tons per year, from Lester R. Brown et al., *Vital Signs* (New York: W.W. Norton & Co., 1993).

10. Laureti, op. cit., note 9.

11. FAO, *Marine Fisheries*, op. cit., note 5. More in-depth subsidy discussion comes later.

12. Ibid.

13. Job data in Table 4. See note 77.

14. There is evidence that Neanderthals (40,000 to 50,000 years ago) caught fresh-water fish, probably without tools, Cro Magnon (25,000 to 30,000 years ago) may have used some fishing tools, and by 15,000 to 8,000 B.C., Magdalenian culture in western Europe had well-developed fishing technology, including barbed hooks and traps, from Brian M. Fagan, *People of the Earth: An Introduction to World Prehistory* (Boston: Little, Brown and Company, 1983); Baltic from James R. McGoodwin, *Crisis in the World's Fisheries* (Stanford, California: Stanford University Press, 1990).

15. The 9.9 percent of the ocean area that lies over continental shelves, and the 0.1 percent of the oceans that include upwelling zones are the primary fisheries of the world, from John Ryther, "Photosynthesis and Fish Production in the Sea," *Science*, Vol. 166, 1969. Fishing grounds corresponding with areas of marine productivity from U.S. Central Intelligence Agency, *The New Global Fishing Regime: Impact and Response* (Washington, D.C.: General Printing Office, June 1980). For a general discussion, see Peter Weber, *Abandoned Seas: Reversing the Decline of the Oceans*, Worldwatch Paper 116 (Washington, D.C.: Worldwatch Institute, November 1993). Around the Atlantic, the major fishing grounds include the Grand Banks, Europe's North Sea, the Mediterranean Sea, the continental shelf off western Africa, the Benguela current off the southwest coast of Africa, and the Patagonia Shelf off Argentina. The Pacific, the Peruvian upwelling, the Bering Sea, Russia's Sea of Okhotsk, and Japan's Kuroshio and Oyashio currents are among the most productive fisheries in the world. The waters of Southeast Asia, the Bay of Bengal, and the Arabian Sea also support rich fisheries.

16. For a discussion of productivity and food potential, see Frederick W. Bell, *Food From the Sea: The Economics and Politics of Ocean Fisheries* (Boulder, Colorado: Westview Press, 1978).

17. Ibid.

18. 1968 conference results from Michael L. Weber, Washington, D.C., private communication, May 31, 1994; 100 million ton estimate from the FAO-sponsored publication, J.A. Gulland, ed., *The Fish Resources of the Ocean* (Surrey, England: Fishing News Ltd., 1971). This estimate is meant to include traditional bony fish ranging from commonly eaten species such as cod and haddock to the small shoaling species such as the Peruvian anchovy. The range from M.A. Robinson, *Trends and Prospects in World Fisheries*, Fisheries Circular No. 772 (Rome: FAO, 1984) cited in World Resources Institute (WRI), *World Resources 1992-93* (New York: Oxford University Press, 1992). Robinson's estimate includes crustaceans and cephalopods (technically a type of mollusk) along with traditional bony fish.

19. Current projections from FAO, *Marine Fisheries*, op. cit., note 5.

20. "Catch" in this paper refers to wild catch only; FAO "catch" statistics include aquaculture yields as well. Estimate for 1900 from Ray Hilborn, "Marine Biota" in B.L. Turner et al., eds., *The Earth as Transformed by Human Action* (New York: Cambridge University Press, 1990); growth rates from FAO, *Marine Fisheries*, op. cit., note 5; 1984-91 aquaculture, inland wild and marine wild data from Maurizio Perotti, FAO, Fishery Information, Data and Statistics Service (FIDI), unpublished printout, November 3, 1993. 1950-91 world catch, marine catch, and inland catch data from *FAO Yearbook of Fishery Statistics: Catches and Landings*, (Rome: 1967-91); before 1984, estimates are based on a 1975 aquaculture production estimate from National Research Council, Aquaculture in the United States: Constraints and Opportunities, (Washington, D.C.: National Academy of Sciences, 1978) and country estimates in Conner Bailey and Mike Skladeny, "Aquacultural Development in Tropical Asia," *Natural Resources Forum*, February 1991.

21. D. Pauly and I. Tsukayama, *The Peruvian Anchoveta and Its Upwelling Ecosystem: Three Decades of Change* (Manila: International Center for Living Aquatic Resources Management, 1987).

22. FISHSTAT-PC, op. cit., note 4.

23. Ibid.

24. Ibid.

25. Ibid.

26. Ibid.

27. Ibid and *FAO Yearbook of Fishery Statistics: Catches and Landings*, (Rome: FAO, various years pre-1970s.)

28. FAO, *Marine Fisheries*, op. cit., note 5.

29. Ibid.

30. Production of carp from FISHSTAT-PC, op. cit., note 4. See endnote 20 for aquaculture estimates.

31. Examples in Hilborn, op. cit., note 20; and McGoodwin, op. cit., note 14.

32. El Niño described in J.T. Houghton, G.J. Jenkins, and J.J. Ephraums, eds., *Climate Change: The IPCC Scientific Assessment* (New York: Cambridge University Press, 1990); early Peru from McGoodwin, op. cit., note 14 and Fagan, op. cit., note 14.

33. Discussion of Peruvian anchovy in Pauly and Tsukayama, op. cit., note 21 and William E. Evans, "Management of Large Marine Ecosystems" in Kenneth Sherman and Lewis M. Alexander, eds., *Biomass Yields and Geography of Large Marine Ecosystems* (Boulder, Colorado: Westview Press, 1989). The Sherman and Alexander book provides excellent case studies of large ecosystem biomass shifts and their causes.

34. Depleted species from John Caddy, FAO Fisheries Division, private communication, March 23, 1994; depleted coastal waters from FAO, *Marine Fisheries*, op. cit., note 5.

35. Definitions of types of overfishing from John P. Wise, *Federal Conservation & Management of Marine Fisheries in the United States* (Washington, D.C.: Center for Marine Conservation, 1991).

36. "Overfishing Depletes Exotic Southern Ocean Fish," *Christian Science Monitor*, August 14, 1991.

37. Massachusetts Offshore Groundfish Task Force, *New England Groundfish in Crisis—Again* (Boston, Massachusetts: Executive Office of Environmental Affairs, 1990) cited in World Resources Institute, op. cit., note 18.

38. Penelope Cumler, Portland, Maine, private communication, June 13, 1994.

39. Chris Newton, FAO Fisheries Division, private communication, March 25, 1994. Because bycatch is undocumented, estimates vary widely. The total bycatch from shrimp in developing countries is estimated to be 10 to 15 million tons per year in Basil Hinds, "The Economics of Fish Resources in Lesser Developed Countries: The Fish By-Catch and Externalities in Shrimp Fishery," Ph.D. Dissertation, Howard University, 1981 cited in Hinds, op. cit., note 7. Shrimp bycatch estimated at 4.5 to 19 million tons per year with 1/4 to 1/2 discarded; the total bycatch is estimated at 9 million tons per year in Eugene C. Bricklemyer, Jr., Suzanne Iudicello, and Hans J. Hartmann, "Discarded Catch in U.S. Commercial Marine Fisheries" in William J. Chandler, *Audubon Wildlife Report 1989/1990* (New York: Academic Press, Inc., 1989).

40. Shrimp and red snapper from *Our Living Oceans: Report on the Status of U.S. Living Marine Resources, 1993* (Silver Spring, Maryland: National Marine Fisheries Service, December 1993). See endnote 39 for discussion of bycatch data. FAO estimates that 80 to 90 percent of the shrimp catch in the tropics is bycatch in FAO, *Marine Fisheries*, op. cit., note 5.

41. Han J. Lindeboom, "How Trawlers are Raking the North Sea to Death," *The Daily Telegraph*, March 19, 1990.

42. Driftnet bycatch from Andy Palmer, American Oceans Campaign, private communication, September 27, 1993; purse seine dolphin catch from Eric Christensen and Samantha Geffin, "GATT Sets Its Nets on Environmental Regulation: The GATT Panel Ruling on Mexican Yellowfin Tuna Imports and the Need for Reform of the International Trading System," *The University of Miami Inter-American Law Review*, Winter 1991-92 and Hilary F. French, "The Tuna Test: GATT and the Environment," *World Watch*, March/April 1992.

43. Greenpeace International, *It Can't Go On Forever: The Implications of the Global Grab for Declining Fish Stocks* Amsterdam, July 1993.

44. Steller sea lion from *It Can't Go On Forever*, op. cit., note 43; pollock from Natalia S. Mirovitskaya and J. Christopher Haney, "Fisheries Exploitation as a Threat to Environmental Security: The North Pacific Ocean," *Marine Policy*, July 1992.

45. No systematic global estimates for loss due to pollution and habitat destruction exist because of the difficulty of attributing cause. Nonetheless, we know that salmon losses total at least 500,000 tons per year; losses in such semi-enclosed seas and estuaries as the Baltic, the Chesapeake, the Yellow Sea, the Black Sea, Azov Sea, and the Aral Sea indicate that other losses are at least on the order of 500,000 tons per year; coastal wetland loss may have reduced productivity on the order of 4 million tons per year; and coral reef destruction, another 500,000 tons per year, based on various sources included in text.

46. Salmon from Hilborn, op. cit., note 20; former Soviet Union water diversion from Michael A. Rozengurt, "Alternation of Freshwater Inflows" in Richard H. Stroud, ed., *Stemming the Tide of Coastal Fish Habitat Loss*, Proceedings of a Symposium on Conservation of Coastal Fish Habitat, Baltimore, Maryland, March 7-9, 1991 (Savannah, Georgia: National Coalition for Marine Conservation, 1992).

47. Aswan dam discussion in Adam Ben-Tuvia, "The Mediterranean Sea, Biological Aspects" in *Ecosystems of the World, Estuaries and Enclosed Seas*, Vol. 26 (New York: Elsevier Scientific Publishing Company, 1983).

48. Total and mangrove wetland loss from WRI, op. cit., note 18; Indonesia example and estimate of 480 kilograms per hectare per year of lost catch from mangroves from Nora Berwick, "Background Paper On Indonesia's Coastal Resources Sector," Bureau for Science and Technology, U.S. Agency for International Development, Washington, D.C., April 1989; Ecuador example from Silvia Q. Coello, "Mangrove Swamps Disappearing," *El Universo*, December 5, 1993 reported in *Environmental Issues*, February 7, 1994. Estimate of mangrove loss, from WRI list of selected countries, is 9.8 million hectares; total loss is higher. Estimate of 160 kilograms of shrimp per hectare per year from Lawrence S. Hamilton and Samuel C. Snedaker, eds., *Handbook for Mangrove Area Management*, collaboration of East-West Center, International Union for the Conservation of Nature and Natural Resources, and United Nations Educational, Scientific and Cultural Organization, with the United Nations Environment Programme (Honolulu, Hawaii: East-West Environmental and Policy Center, 1984).

49. Calculation for coral reefs based on 5-10 percent loss from Clive R. Wilkinson, "Coral Reefs are Facing Widespread Extinctions: Can We Prevent These Through Sustainable Management Practices?" presented at the Seventh International Coral Reef Symposium, Guam, 1992; 600,000 square kilometers of reef from S.V. Smith, "Coral-Reef Area and the Contributions of Reefs to Processes and Resources of the World's Oceans," *Nature*, May 18, 1978; productivity ranging from 8 to 35 tons square kilometers per year from International Center for Living Aquatic Resources Management (ICLARM), *ICLARM's Strategy For International Research on Living Aquatic Resources Management* (Manila, Philippines: 1992). These figures yield a range of 240,000 tons/yr (on potential 4.8 mmt/yr) to 2.1 mmt/yr (on potential 21 mmt/yr) in lost potential. Estimated world catch of 4 mmt/yr by 4 million fishers (ICLARM and J. Caddy, Fisheries Department, FAO, Rome, private communication, September 25, 1992). Best estimate using low value: 0.25-0.5 mmt/yr. This calculation reflects the decline in reef fish; silted and over-grown reefs can actually support higher catches of other fish.

50. China from Zhijie and Côté, op. cit., note 6; pollution and oyster catch from *The Chesapeake Bay: A Progress Report, 1990-91*, Chesapeake Executive Council,

Annapolis, Maryland, August 1991; individual declines from James R. Chambers, "Coastal Degradation and Fish Population Losses" in Stroud, ed., op. cit., note 46; filter example from Palmer, op. cit., note 42.

51. John Cairns, Jr. and James R. Pratt, "Aquatic Toxicology - Ninth Volume," *Factors Affecting the Acceptance and Rejection of Genetically Altered Microorganisms by Established Natural Aquatic Communities,* Special Technical Publication 921 (Philadelphia, Pennsylvania: American Society for Testing and Materials, 1986); John Travis, "Invader Threatens Black, Azov Seas," *Science,* November 26, 1993; John Pomfret, "Black Sea, Strangled by Pollution, Is Near Ecological Death," *Washington Post,* June 20, 1994.

52. Suez canal example from Elliot A. Norse, ed., *Global Marine Biological Diversity: A Strategy for Building Conservation into Decision Making* (Washington, D.C.: Island Press, 1993); other examples from Cairns and Pratt, op. cit., note 51 and Eric Slaughter, U.S. Environmental Protection Agency, National Estuary Program, private communication, March 17, 1994; species in transit from James T. Carlton and Jonathan B. Geller, "Ecological Roulette: The Global Transport of Nonindigenous Marine Organisms," *Science,* July 2, 1993.

53. *The Environment in Sweden-Status and Trends: Environmental Pollution and Health,* Report No. 4249 (Solna, Sweden: Swedish Environmental Protection Agency, October 1993).

54. For a complete discussion of this topic see Ann Misch, "Assessing Environmental Health Risks," *State of the World 1994* (New York: W.W. Norton & Co., 1994.) See also: Theo Colborn, Frederick vom Saal, and Ana Soto, "Developmental Effects of Endocrine-Disrupting Chemicals in Wildlife and Humans," *Environmental Health Perspectives,* Vol. 101, No. 5 (Research Triangle Park, North Carolina: National Institute of Environmental Health Sciences, 1993); Stephen Bortone and William Davis, "Fish Intersexuality as Indicator of Environmental Stress," *BioScience,* March 1994.

55. Eric Dewailly, Pierre Ayotte, Suzanne Bruneau, Claire Laliberte, Derek Muir, and Ross Nortsrom, "Innuit Exposure to Organochlorines through the Aquatic Food Chain in Arctic Quebec," *Environmental Health Perspectives,* Vol. 101, No. 7, December 1993; Health effects from National Research Council, *Environmental Epidemiology: Public Health and Hazardous Wastes* (Washington, D.C.: National Academy Press, 1991).

56. Christopher Anderson, "Cholera Epidemic Traced to Risk Miscalculation," *Nature,* November 28, 1991.

57. United States from Farid E. Ahmed, ed. *Seafood Safety,* (Washington, D.C.: National Academy Press, 1991); hospitalizations from Martin W. Holdgate, "The Sustainable Use of Tropical Coastal Resources: A Key Conservation Issue," *Ambio,* November 1993.

58. Health standards from David Bailey, Senior Attorney, Environmental Defense Fund, private communication, June 27, 1994; subsistence fishers in Ahmed, ed., op. cit., note 57. See also: Environmental Defense Fund, Legal Department of the National Association for the Advancement of Colored People, and Penobscot Indian Nation, "The Protection of Sport and Subsistence Fishing

Populations in the United States," Recommendations to the Administrator of the U.S. Environmental Protection Agency for Implementation of the President's Executive Order on Environmental Justice and the Subsistence Consumption of Fish and Wildlife, Washington, D.C., June 1994.

59. See for example "Mediterranean Cetaceans Face Deadly Pollution," *Journal of Commerce*, April 21, 1992; J.R. Geraci, "Clinical Investigation of the 1987-88 Mass Mortality of Bottlenose Dolphins Along the U.S. Central and South Atlantic Coast," final report to the National Marine Fisheries Service, U.S. Navy Office of Naval Research, and Marine Mammal Commission, Washington, D.C., April 1989; and Greenpeace, "Critique of the National Oceanic and Atmospheric Administration's Final Report on the Clinical Investigation of the 1987-88 Mass Mortality of Bottlenose Dolphins Along the U.S. Central and South Atlantic Coast," Washington, D.C., May 1989.

60. Scientific consensus predictions for global warming from Houghton, et al., *Climate Change*, op. cit., note 32, updated in IPCC, *Climate Change 1992: The IPCC Supplementary Report* (New York: Cambridge University Press, 1992); Little Ice Age from Walter Sullivan, "Study of Greenland Ice Finds Rapid Change in Past Climates," *New York Times*, July 15, 1993 and Kathy Sawyer, "Climatology: Warming Could Trigger Cold Spells," *Washington Post*, July 19, 1993. For more discussion of the potential effects of global warming on marine fisheries see Robert C. Francis, "Climate Change and Marine Fisheries" and Victor S. Kennedy, "Anticipated Effects of Climate Change on Estuarine and Coastal Fisheries" in *Fisheries*, November-December 1990, and P.A. Fields et al., "Effects of Expected Global Climate Change on Marine Faunas," *TREE*, Elsevier Science Publishers Ltd., London, October 1993 (reprint from California Sea Grant).

61. McGoodwin, op. cit., note 14.

62. Refer to sources for Table 4, note 77.

63. Boat numbers from Susan Pollack, "No More Fish Stories," *The Amicus Journal*, Spring 1994; Newton, op. cit., note 39.

64. Norway and Europe from Carl-Christian Schmidt, "The Net Effects of Over-fishing," *The OECD Observer*, October/November 1993.

65. Bonnie McCay et al., "Privatization in Fisheries: Lessons from Experiences in the U.S., Canada, and Norway," presented at Symposium of the Ocean Governance Study Group: "Moving Ahead on Ocean Governance: Practical Applications Guided by Long-Range Visions," Lewes, Delaware, April 9-13, revised March 31, 1994.

66. Bill Shapiro, "The Most Dangerous Job in America," *Fortune*, May 31, 1993.

67. Don Hinrichsen, presentation at Worldwatch Institute, July 16, 1993; world population in coastal cities from WRI, op. cit., note 18; and Eugene Robinson, "Worldwide Migration Nears Crisis," *Washington Post*, July 7, 1993.

68. Subsidies from FAO, *Marine Fisheries*, op. cit., note 5; Malaysia policy from James Clad, "The Fish Catches it," *Far Eastern Economic Review*, June 21, 1984; Malaysia population from United Nations Department for Economic and Social Information and Policy Analysis (UNDESIPA), *World Population Prospects: The*

1992 Revision (New York: 1993); Malaysia exchange rate of M$2.2 per U.S.$1 from International Monetary Fund, *International Financial Statistics Yearbook* (Washington, D.C.: 1993).

69. Taiwan from Julian Baum, "Drifting Downstream," *Far Eastern Economic Review*, August 29, 1991; Russia from Newton, op. cit., note 39; estimate for the United States from Mike Weber, private communication, June 1, 1994.

70. John Kurien and T.R. Thankappan Achari, "Overfishing along Kerala Coast: Causes and Consequences," *Economic and Political Weekly*, September 1-8, 1990.

71. Ibid.

72. Ibid.

73. Eduardo A. Loayza and Lucian M. Sprague, *A Strategy for Fisheries Development*, World Bank Discussion Papers, Fisheries Series, No. 135 (Washington, D.C.: The World Bank, 1992).

74. Vlad Kaczynski, School of Marine Affairs, University of Washington, Seattle, Washington, private communication, April 14, 1994.

75. Alfredo Sfeir-Younis and Graham Donaldson, *Fishery Sector Policy Paper*, (Washington, D.C.: The World Bank, December 1982), and Loayza and Sprague, op. cit., note 73.

76 Fishers in Japan and China from Adele Crispoli, Fishery Statistician, FAO, Fishery Information, Data and Statistics Service (FIDI), private communication, April 8, 1994.

77. Table 4 adapted from Conner Bailey, "Optimal Development of Third World Fisheries" in Michael A. Morris, ed., *North-South Perspectives on Marine Policy* (Boulder, Colorado: Westview Press, 1988). Employment data from FAO, *Marine Fisheries*, op. cit., note 5 with additional estimates for small-scale fishers from International Center for Living Aquatic Resources Management (ICLARM), *ICLARM 1992 Report* (Manila, Philippines: 1993); R.S. Pomeroy and A. Cruz-Trinidad, "Socio-economic Aspects of Artisanal Fisheries in Asia" in S.S. de Silva, ed., *Asian Fisheries Society Commemorative Volume* (Manila, Philippines: Asian Fishery Society, in press); and H. Josupeit, *The Economic and Social Effects of the Fishing Industry: A Comparative Study*, FAO Fisheries Circular No. 314 (Rome: 1981). Income, fuel consumption, and investment figures from *Marine Fisheries*. Catch estimates based on *Marine Fisheries*.

78. Income estimates from FAO, *Marine Fisheries*, op. cit., note 5; Newfoundland from Peter R. Sinclair, "Introduction" in Peter R. Sinclair, ed., *A Question of Survival* (St. John's, Newfoundland: Institute of Social and Economic Research, 1988); Asia and India figures from Pomeroy and Cruz-Trinidad, op. cit., note 77. For further discussion of fisher culture, see McGoodwin, op. cit., note 14.

79. Mediterranean and Persian Gulf examples from John Caddy, op. cit., note 34.

80. Taiwan from Baum, op. cit., note 69; Malaysia from Robert Birsel, "Malaysian Fishermen Under Threat," *Pakistan & Gulf Economist*, August 1-7, 1987 and Mohamed, op. cit., note 7; Europe from Commission of the European Communities, *Report 1991 from the Commission to the Council and the European*

Parliament on the Common Fisheries Policy, Brussels, December 18, 1991; 40 percent from Schmidt, op. cit., note 64; Japanese buy-backs from Alan Macnow, Federation of Japan Tuna Fisheries Cooperative Associations, New York, private communication, April 15, 1994; Japan north Pacific reductions from Olav Schram Stokke, "Transnational Fishing: Japan's Changing Strategy," *Marine Policy*, July 1991.

81. McCay et al., op. cit., note 65.

82. Ibid. McCay did not use the term "windfall".

83. New Zealand example from Mark Feldman, "Fishing Boom, Fishing Bust, a Cautionary Tale," *Forest & Bird*, May 1990 and Mark Bellingham, "A Better Deal for Life in the Sea?" *Forest & Bird*, February 1993.

84. Linda Binken, Alaska Longline Fishermen's Association, Sitka, Alaska, private communication, April 2, 1994.

85. Population growth rate and Mexican population from Population Reference Bureau, *1993 World Population Data Sheet* (Washington, D.C.: 1993); population projections from UNDESIPA, op. cit., note 68. The necessary increases are calculated on the basis of the fish supply that goes for human consumption, 70 million tons in 1990, not the portion that goes to animal feed. Maintaining the status quo for animal feed would require even greater increases, which are not likely with the species currently used for this purpose.

86. Rehabilitation from FAO, *Marine Fisheries*, op. cit., note 5; for aquaculture growth see endnote 20. Projecting future growth rates is dubious, especially for a nascent industry such as aquaculture; no prediction is intended by the given figures. This is only one possible scenario.

87. Laureti; op. cit., note 9.

88. Protein figures from Laureti, op. cit., note 9 and *FAO Yearbook of Fishery Statistics: Commodities* (Rome: FAO, 1993).

89. For Figure 2, beef, pork and chicken prices from *FAO Yearbook of Production: Trade and Commerce* (Rome: various years) and fish prices from *FAO Yearbook of Fishery Statistics: Commodities* (Rome: various years). See price discussion in FAO, *Marine Fisheries*, op. cit., note 5.

90. Kurien and Achari, op. cit., note 70; conversion to U.S. dollars based on 1961 and 1981 exchange rates, 4.765 and 10.591 rupees per dollar respectively, from International Monetary Fund, *International Financial Statistics* (Washington, D.C.: 1990).

91. According to "Vietnam to Improve Quality of Its Seafood," *Journal of Commerce*, March 18, 1994, Denmark and U.N. Industrial Development Organization are signing a $1.6 million agreement to improve quality and help the country achieve its goal of doubling exports to $0.9 to 1 billion by 2000.

92. McGoodwin, op. cit., note 14.

93. Michael Gawel, Territorial Planning Office of Guam, private communication, August 20, 1992.

94. FAO, *Marine Fisheries*, op. cit., note 5; European Parliament, Directorate General for Research, *European Community Fisheries Agreements with Third Countries and Participation in International Fisheries Agreements* (Luxembourg: 1993); percentage catch from FISHSTAT-PC, op. cit., note 4.

95. Kaczynski, op. cit., note 74.

96. *FAO Yearbook of Fisheries Statistics: Commodities* (Rome: 1993). In 1991, these countries accounted for a total of $36.6 million worth of world fisheries imports.

97. Ibid; Stokke, op. cit., note 80.

98. Developing country populations from UNDESIPA, op. cit., note 68; fish supply from Laureti, op. cit., note 9.

99. *FAO Yearbook of Fishery Statistics: Commodities*, and *FAO Yearbook of Fishery Statistics: Catches and Landings* (Rome: 1993).

100. Chile from John F. Kearney, "Restoring the Common Wealth of Ocean Fisheries," A Discussion Paper Oriented toward Enlarging the Concept of Sustainability in the Deliberations Leading to the UN Conference On Straddling and Highly Migratory Fish Stocks, prepared for The Conservation Council of New Brunswick, June 1993.

101. Ibid.

102. Developing country populations from U.N., *World Population Prospects*, op. cit., note 68; fish supply from Laureti, op. cit., note 9.

103. China exports from *FAO Yearbook of Fishery Statistics: Commodities*, op. cit., note 96.

104. Aquaculture data from sources in endnote 20.

105. FAO, *Marine Fisheries*, op. cit., note 5.

106. Wolfgang Hannig, "Innovation and Tenant Survival: Brackish Pond Culture in Java," *NAGA, The ICLARM Quarterly*, April 1988.

107. Chris Wille, "The Shrimp Trade Boils Over," *International Wildlife*, November/December 1993.

108. See for example, J. Honculada Primavera, "A Critical Review of Shrimp Pond Culture in the Philippines," *Reviews in Fisheries Science*, Volume 1, Number 2, 1993; *Wild Salmon: Present and Future*, proceedings of international conference, Sherkin Island Marine Station, Sherkin Island, Ireland, September 16-17, 1988; and overview from Hal Kane, "Growing Fish in Fields," *World Watch*, September/October 1993.

109. Ghana from *NAGA ICLARM Quarterly*, April-July 1993, photosection; C. Kwei Lin and Christopher Lee, "Production of Freshwater Prawns in the Mekong Delta," *NAGA ICLARM Quarterly*, April 1992; "Shrimp Farming in Indonesia," *Appropriate Technology International*, Bulletin No. 20 (Washington, D.C.: Appropriate Technology International, November 1989).

110. Paul Skillicorn, William Spira, and William Journey, *Duckweed Aquaculture: A New Aquatic Farming System for Developing Countries* (Washington, D.C.: World Bank, 1993) and Paul Skillicorn, World Bank, private communication, November 14, 1993.

111. *Mare Liberum* argument in FAO, *Marine Fisheries*, op. cit., note 5. Hugo Grotius, *Mare Liberum (The Freedom of the Seas)* (1609), trans., R.V.D. Magoffin, (New York: Oxford University Press, 1916) cited in McGoodwin, op. cit., note 14. One nautical mile is equal to 1.15 English miles or 1.85 kilometers.

112. *The Law of the Sea: United Nations Convention on the Law of the Sea* (New York: United Nations, 1983); Judith Fenwick, *International Profiles on Marine Scientific Research* (Woods Hole, Massachusetts: Woods Hole Oceanographic Institution, 1992).

113. Ibid.

114. "South Pacific Forum: Final Act of the Meeting on a Convention to Prohibit Driftnet Fishing in the South Pacific, Including Text of Convention for the Prohibition of Fishing with Long Driftnets in the South Pacific and its Protocols (November 24, 1989)," *International Legal Materials*, November 1990; first U.N. resolution against driftnets from "United Nations: General Assembly Resolution on *Large-Scale Pelagic Driftnet Fishing and Its Impact on Living Marine Resources of the World's Oceans and Seas* (passed December 22, 1989)," *International Legal Materials*, November 1990; 1990 and 1991 U.N. resolutions from Mike Hagler, Greenpeace International, Auckland, New Zealand, private communication, October 1, 1993.

115. Falklands from Stephen Fidler and John Barham, "Flying into Flak over Fish and Falklands," *Financial Times*, January 6, 1993 and John Barham, "Argentina Secures Bigger Share of Fishery Resources," *Financial Times*, October 21, 1993; South Georgia and South Sandwich Islands from Argentine and Great Britain government statements in *Law of the Sea Bulletin*, December 1993. 30 million British pounds converted to U.S. dollars based on 1.48 pounds per dollar in Foreign Exchange listing in *Washington Post*, March 9, 1994.

116. Nickerson, op. cit., note 1. For more information on regional fishing agreements, see Michael J. Savini, *Summary Information on the Role of International Fishery Bodies with Regard to the Conservation and Management of Living Resources of the High Seas*, FAO Fisheries Circular No. 835, Revision 1 (Rome: FAO, September 1991).

117. "Bluefin Tuna Reported on Brink of Extinction," *Journal of Commerce*, November 10, 1993. General discussion from Carl Safina, "Bluefin Tuna in the West Atlantic: Negligent Management and the Making of an Endangered Species," *Conservation Biology*, June 1993; restriction on catch from Debora MacKenzie, "Too Little Too Late to Save Atlantic Bluefin," *New Scientist*, November 20, 1993.

118. Non-ICCAT catch and trade debate from Safina, op. cit., note 117; non-ICCAT vessels from Newton, op. cit., note 39.

119. Overview in Colin Nickerson, "Stripping the Sea's Life," *Boston Sunday Globe*, April 17, 1994. See text for Northwest Atlantic, Atlantic Tuna, and

Patagonia Shelf; Northeast Atlantic from Motoko Rich, "Icelandic Fishing Boat Seized in Atlantic," *Financial Times*, March 22, 1994, Alan Riding, "Fishermen Take Arms Against a Sea of Troubles," *New York Times*, April 1, 1993; David Gardner, "EU Entry Talks Struggle to Find Accord on Fishing," *Financial Times*, February 22, 1994, and International Court of Justice, "Maritime Delimitation in the Area between Greenland and Jan Mayen (Denmark v. Norway)," Judgment of the Court, *Law of the Sea Bulletin*, December 1993; West African example of foreign fishers depleting traditional fisheries from J.M. Vakily, "Assessing and Managing the Marine Fish Resources of Sierra Leone, West Africa," *Naga*, The ICLARM Quarterly, Manila, January 1992; Namibia from Asser Ntinda, "Fishing Pirates Stir Up Namibian Waters," *Panoscope*, July 1991 and Asser Ntinda, "Namibia Halts the Plunder," *Panoscope*, November 1992; Southeast Asia from Rizal Sukma, "Indonesia and the South China Sea: Interests and Policies," *The Indonesian Quarterly*, Volume 20, Number 4 and George Kent and Mark J. Valencia, eds., *Marine Policy in Southeast Asia* (Los Angeles: University of California Press, 1985); "Curb on Bering Sea Fishing Tentatively Set," *New York Times*, February 14, 1994; Peanut Hole discussed in Viktor Yurlov, "Under the 'Flag of Convenience,'" *Robochaya Tribuna*, Moscow, March 3, 1993, reprinted in *Environmental Issues*, Washington, D.C., Foreign Broadcast Information Service, March 31, 1993; driftnetting status from Hagler, op. cit., note 114; whaling update from Bronwen Maddox, "World's Whales Worth More Alive than Dead," *Financial Times*, May 16, 1994.

120. Development of the split from David E. Pitt, "U.N. Talks Combat Threat to Fishery," *New York Times*, July 25, 1993 and Palmer, op. cit., note 42. Additional information on the U.N. Conference on Straddling and Highly Migratory Fish Stocks is available in *Earth Negotiations Bulletin*, December 21, 1993 and *ECO*, the March 1994 issues, published by Greenpeace, Washington, D.C.

121. Sealing treaty from FAO, *Marine Fisheries*, op. cit., note 5.

122. *The Law of the Sea*, op. cit., note 112.

123. Newton, op. cit., note 39.

124. FAO, *Marine Fisheries*, op. cit., note 5.

125. Ross, op. cit., note 6.

126. Discussion of traditional management in Gary A. Klee, "Oceania," in Gary A. Klee, ed., *World Systems of Traditional Resource Management* (New York: John Wiley & Sons, 1980); R.E. Johannes, CSIRO Marine Laboratories, Australia, "Small-Scale Fisheries: A Storehouse of Knowledge for Managing Coastal Marine Resources," presented at Ocean Management Symposium, Smithsonian Institution, Washington, D.C., November 20, 1991; Conner Bailey and Charles Zerner, "Role of Traditional Fisheries Resource Management Systems for Sustainable Resource Utilization," presented at Perikanan Dalam Pembangunan Jangka Panjang Tahap II: Tantangan dan Peluang, Sukabumi, West Java, June 18-21, 1991.

127. Maine from James M. Acheson, "The Lobster Fiefs Revisited: Economic and Ecological Effects of Territoriality in Maine Lobster Fishing," in Bonnie J. McCay and James M. Acheson, eds., *The Question of the Commons: The Culture and*

Ecology of Communal Resources (Tucson, Arizona: The University of Arizona Press, 1987).

128. Kenneth Ruddle, "The Continuity of Traditional Management Practices: The Case of Japanese Coastal Fisheries," in Kenneth Ruddle and R.E. Johannes, eds., *The Traditional Knowledge and Management of Coastal Systems in Asia and the Pacific* (Jakarta Pusat, Indonesia: UNESCO, 1985).

129. Ibid.

130. Charles Zerner, *Imagining Marine Resource Management Institutions in the Maluka Islands, Indonesia 1870-1992*, Case Study No. 6, prepared for the Liz Claiborne and Art Ortenberg Foundation Community Based Conservation Workshop, Airlie, Virginia, October 18-22, 1993.

131. Philippine example from Hinrichsen, op. cit., note 67; see also, Don Hinrichsen, "Philippine Mangroves: Bounty in the Brine," *International Wildlife*, May/June 1992 and "Managing Mangroves in the Philippines," *People*, November 3, 1991.

132. Eduardo A Loayza, ed., *Managing Fishery Resources*, Proceedings of a Symposium Co-sponsored by the World Bank and Peruvian Ministry of Fisheries, held in Lima, Peru, June 1992, World Bank Discussion Papers, Fisheries Series No. 217 (Washington, D.C.: 1994); The Gambia from Fishery Committee for the Eastern Central Atlantic, "Report of the Ninth Session of the Working Party on Resource Education," Lagos, Nigeria, November 19-23, 1990 (Rome: FAO, 1992).

133. Vakily, op. cit., note 119.

134. FAO, *Marine Fisheries*, op. cit., note 5.

135. Europe and Iceland from Schmidt, op. cit., note 64.

136. Australia rents from FAO, *Marine Fisheries*, op. cit., note 5; quota system percentage from "Fish: The Tragedy of the Oceans," *The Economist*, March 19, 1994 and public information brochure, New South Wales Fisheries, Pyrmont, Australia.

137. A discussion of U.S. fishery law is in Wise, op. cit., note 35.

138. Kaczynski, op. cit., note 74.

139. Nordmore grate example from David Allison, Washington, D.C., private communication, May 2, 1994.

140. Earl Krygir, Alaska Department of Fish and Game, Juneau, Alaska, private communications, May 1994; Alaska Marine Conservation Council, Anchorage, Alaska, Harvest Priority fact sheet.

141. Krygir, op. cit., note 140.

142. For a more extensive discussion of the threats to the oceans, see Weber, op. cit., note 15.

143. For example of losses from logging, see Gregor Hodgson and John A. Dixon, *Logging Versus Fisheries and Tourism in Palawan: An Environmental and Economic Analysis*, Occasional Paper No. 7 (Honolulu, Hawaii: East-West Environmental and Policy Center, 1988).

144. Verhovek, op. cit., note 3.

145. Japan from Ruddle, op. cit., note 128.

146. Dan Seligman, Sierra Club, Washington, D.C., private communication, March 9, 1994.

THE WORLDWATCH PAPER SERIES

No. of
Copies

_____ **Total Copies**

☐ **Single Copy: $5.00**
☐ **Bulk Copies (any combination of titles)**
　　☐ 2–5: $4.00 ea.　　☐ 6–20: $3.00 ea.　　☐ 21 or more: $2.00 ea.

☐ **Membership in the Worldwatch Library: $30.00 (international airmail $45.00)**
The paperback edition of our 250-page "annual physical of the planet," *State of the World 1994*, plus all Worldwatch Papers released during the calendar year.

☐ **Subscription to *World Watch* Magazine: $20.00 (international airmail $35.00)**
Stay abreast of global environmental trends and issues with our award-winning, eminently readable bimonthly magazine.

Please include $3 postage and handling per order.

Make check payable to Worldwatch Institute
1776 Massachusetts Avenue, N.W., Washington, D.C. 20036-1904 USA

Enclosed is my check for U.S. $_____

_____

name　　　　　　　　　　　　　　　　**daytime phone #**

_____

address

_____

city　　　　　　　　　　　**state**　　**zip/country**

WWP

THE WORLDWATCH PAPER SERIES

No. of
Copies

_____102. **Women's Reproductive Health: The Silent Emergency** by Jodi L. Jacobson.

_____103. **Taking Stock: Animal Farming and the Environment** by Alan B. Durning and Holly B. Brough.

_____104. **Jobs in a Sustainable Economy** by Michael Renner.

_____105. **Shaping Cities: The Environmental and Human Dimensions** by Marcia D. Lowe.

_____106. **Nuclear Waste: The Problem That Won't Go Away** by Nicholas Lenssen.

_____107. **After the Earth Summit: The Future of Environmental Governance** by Hilary F. French.

_____108. **Life Support: Conserving Biological Diversity** by John C. Ryan.

_____109. **Mining the Earth** by John E. Young.

_____110. **Gender Bias: Roadblock to Sustainable Development** by Jodi L. Jacobson.

_____111. **Empowering Development: The New Energy Equation** by Nicholas Lenssen.

_____112. **Guardians of the Land: Indigenous Peoples and the Health of the Earth** by Alan Thein Durning.

_____113. **Costly Tradeoffs: Reconciling Trade and the Environment** by Hilary F. French.

_____114. **Critical Juncture: The Future of Peacekeeping** by Michael Renner.

_____115. **Global Network: Computers in a Sustainable Society** by John E. Young.

_____116. **Abandoned Seas: Reversing the Decline of the Oceans** by Peter Weber.

_____117. **Saving the Forests: What Will It Take?** by Alan Thein Durning.

_____118. **Back on Track: The Global Rail Revival** by Marcia D. Lowe.

_____119. **Powering the Future: Blueprint for a Sustainable Electricity Industry** by Christopher Flavin and Nicholas Lenssen.

_____120. **Net Loss: Fish, Jobs, and the Marine Environment** by Peter Weber.

_____ **Total Copies**

☐ **Single Copy: $5.00**
☐ **Bulk Copies (any combination of titles)**
   ☐ 2–5: $4.00 ea.   ☐ 6–20: $3.00 ea.   ☐ 21 or more: $2.00 ea.

☐ **Membership in the Worldwatch Library: $30.00 (international airmail $45.00)**
The paperback edition of our 250-page "annual physical of the planet," *State of the World 1994*, plus all Worldwatch Papers released during the calendar year.

☐ **Subscription to *World Watch* Magazine: $20.00 (international airmail $35.00)**
Stay abreast of global environmental trends and issues with our award-winning, eminently readable bimonthly magazine.

Please include $3 postage and handling per order.

Make check payable to Worldwatch Institute
1776 Massachusetts Avenue, N.W., Washington, D.C. 20036-1904 USA

Enclosed is my check for U.S. $_____

_____

name   **daytime phone #**

_____

address

_____

city   **state**   **zip/country**